SpringerBriefs in the Mathematics of Materials

Volume 3

More information about this series at http://www.springer.com/series/13533

Daniel Packwood

Bayesian Optimization for Materials Science

Springer

Daniel Packwood
Institute for Integrated Cell-Materials
Sciences (iCeMS)
Kyoto University
Kyoto
Japan

ISSN 2365-6336 ISSN 2365-6344 (electronic)
SpringerBriefs in the Mathematics of Materials
ISBN 978-981-10-6780-8 ISBN 978-981-10-6781-5 (eBook)
https://doi.org/10.1007/978-981-10-6781-5

Library of Congress Control Number: 2017954484

Printed on acid-free paper

This Springer imprint is published by Springer Nature
The registered company is Springer Nature Singapore Pte Ltd.
The registered company address is: 152 Beach Road, #21-01/04 Gateway East, Singapore 189721, Singapore

Preface

Since the launch of the Materials Genome Initiative in 2011, there has been an increasing interest in the application of statistical and machine-learning techniques to materials science. However, while several high-profile papers have reported the use of such techniques to real problems in materials science, the overall adoption of such techniques by the materials science community remains very limited. A possible cause for this problem is the absence of any textbooks on statistics or machine learning which are specifically aimed at materials scientists.

The purpose of this book is to provide a self-contained tutorial on Bayesian optimization for materials scientists. Bayesian optimization is a machine-learning technique which can enormously accelerate many of the time-consuming tasks in materials science, such as database screening and structure optimization calculations. In Chap. 1, we briefly explain how Bayesian optimization works and give some recent examples of its applications in materials science. In Chap. 2, we provide a self-contained introduction to the theory of Bayesian optimization. This chapter does not assume any advanced mathematical background; however, it does assume that the reader is comfortable with elementary calculus and linear algebra. Upon working through this Chap. 2, the reader should have sufficient knowledge to implement Bayesian optimization into their own research. To help ensure that this is the case, code for performing Bayesian optimization on a simple system is provided (downloadable from the Web) so that the reader see how the theory in the text is implemented in computational setting. Finally, Chap. 3 discusses in detail the application of Bayesian optimization to structure predictions for organic molecules adsorbed to metal surfaces. While the material in this chapter mainly reflects my own research interests, it should nonetheless illustrate how Bayesian optimization is applied to real structure optimization problems.

I wish to thank Masayuki Nakamura and coworkers from Springer Tokyo for helping with the publication of this volume, as well as series editor Prof. Motoko Kotani (Tohoku University) for her tireless efforts to promote mathematics for materials science in Japan. I also thank Prof. Taiji Suzuki (Tokyo University), who introduced me to Bayesian optimization, as well as the members of the Kakanhi Shingakujyutsu project "Exploration of nanostructure-property relationships for materials innovation," who encouraged me to write this book.

Kyoto, Japan
August 2017

Daniel Packwood

Contents

1 Overview of Bayesian Optimization in Materials Science 1
1.1 Brief Overview of Bayesian Optimisation 1
1.2 Examples of Bayesian Optimisation in Materials Science 4
1.2.1 Prediction of Compounds with Low Thermal Conductivity 4
1.2.2 Prediction of Compounds with Optimal Melting Temperatures and Elastic Properties 5
1.2.3 Prediction of Interface Structures 6
1.2.4 Design of Interface Nanostructure 7
1.3 Bayesian Optimization Requires Good Feature Vectors 8
References 10

2 Theory of Bayesian Optimization 11
2.1 Bayesian Interpretation of Probability 11
2.2 Equilibrium Bond Lengths *Via* Bayesian Optimization 12
2.2.1 Prior Probability 14
2.2.2 Likelihood Function and Posterior Distribution 15
2.2.3 Example Calculation of the Posterior Distribution 17
2.2.4 The Expected Improvement 18
2.2.5 Example Run of Bayesian Optimisation 19
2.2.6 Training 22
2.3 Bayesian Optimization in the General Case 23
2.4 *R* Code for Bayesian Optimization. 24
References 28

3 Bayesian Optimization of Molecules Adsorbed to Metal Surfaces 29
3.1 Density Functional Theory for Surface Science 29
3.2 Bayesian Optimization for Surface Science 31

3.2.1 Preliminary Computational Study 32
3.2.2 Statement of Optimization Problem. 34
3.2.3 Data Description 35
3.2.4 Choice of Feature Vectors 36
3.2.5 Training of Hyperparameters 37
3.2.6 Predictive Performance 39
3.2.7 Discussion. 40
References 41

Chapter 1
Overview of Bayesian Optimization in Materials Science

Like any other field of research, materials science involves a lot of trial and error: in the process of creating a new material or device, we will inevitably make several prototypes which fail to perform as hoped. If we could reduce the number of such prototypes, then the materials fabrication process might become faster and less expensive.[1] When President Obama launched the Materials Genome Initiative in 2011, he said the following.

> The invention of silicon circuits and lithium-ion batteries made computers and iPods and iPads possible – but it took years to get those technologies from the drawing board to the marketplace. We can do it faster.
>
> – President Barack Obama, June 2011 [1]

'We can do it faster' refers, in part, to the idea of reducing the number of 'bad' decisions by augmenting the materials fabrication process with clever data analysis. Amongst the various methodologies from statistics and machine learning, Bayesian optimisation shows particular promise for assisting the decision making process in materials science settings. In this Chapter, we will briefly explain how Bayesian optimisation works and discuss several examples of where Bayesian optimisation has been applied in materials science. A technical introduction to Bayesian optimisation is presented in Chap. 2, and an application of Bayesian optimization in surface science will be presented in Chap. 3.

1.1 Brief Overview of Bayesian Optimisation

To introduce Bayesian optimization, we consider the following situation. Consider an alloy of composition M_xN_{1-x}, where M and N are specific metal atoms and x are $1-x$ their respective stoichiometric coefficients. x can take on any value between 0 and 1. Our goal is to find the value of x at which the some property of the alloy (say, hardness) is maximized (Fig. 1.1). One approach might be to create a series of

[1] Of course, prototyping and trial-and-error are necessary for developing scientific understanding.

D. Packwood, *Bayesian Optimization for Materials Science*,
SpringerBriefs in the Mathematics of Materials 3,
https://doi.org/10.1007/978-981-10-6781-5_1

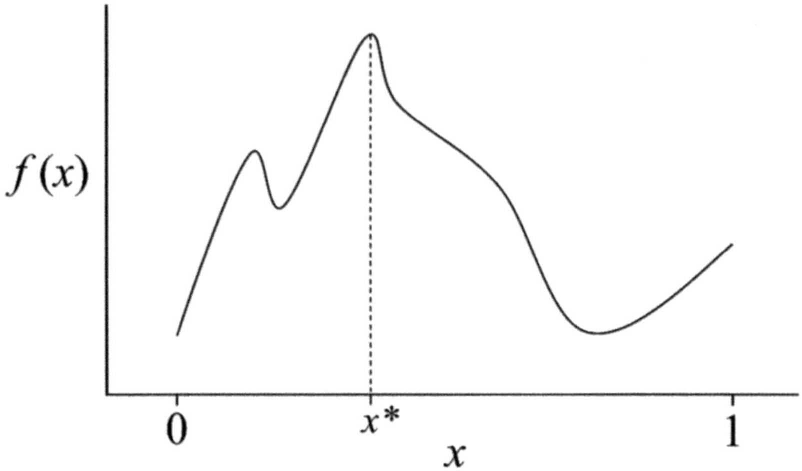

Fig. 1.1 Sketch of the hardness $f(x)$ of alloys of composition M_xN_{1-x}, where M and N represent metal atoms and x represents a stoichiometric coefficient. We can apply Bayesian optimization to find the stoichiometric coefficient x^* at which the hardness is maximized

alloys in which x systematically increases from 0 to 1 and measure the hardness for each one, however this might require a great deal of time and money. We would therefore like to identify the optimal value of x by creating as few alloys as possible.

Let $f(x)$ be the hardness of the alloy M_xN_{1-x}, and let x^* be the value of x which maximizes $f(x)$. In Bayesian optimization, x^* is identified according to the following scheme (Fig. 1.2).

Step 1. Randomly choose n stoichiometric coefficients $x_1, x_2, \ldots,$ and x_n. Create the corresponding alloys and measure their hardness $f(x_1), f(x_2), \ldots, f(x_n)$. The data generated in Step 1 is referred to as the *sample data.*
Step 2. For each possible value of x, assign a probability distribution to the value of the hardness $f(x)$. This distribution is called the *prior probability distribution*. Step 2 is performed independently of Step 1.
Step 3. Using the sample data collected in Step 1, we update the probability distribution constructed in Step 2. This distribution is called the *posterior probability distribution*, and is calculated by applying Bayes' Rule, a formula from probability theory.
Step 4. Let x_m be the stoichiometry of the alloy which has the largest hardness among all alloys in the sample data. With reference to the posterior probability distribution, identify a value of x (outside of the sample data in Step 1) for which the difference

$$f(x) - f(x_m)$$

has a 'high probability' of being maximized (in some sense).

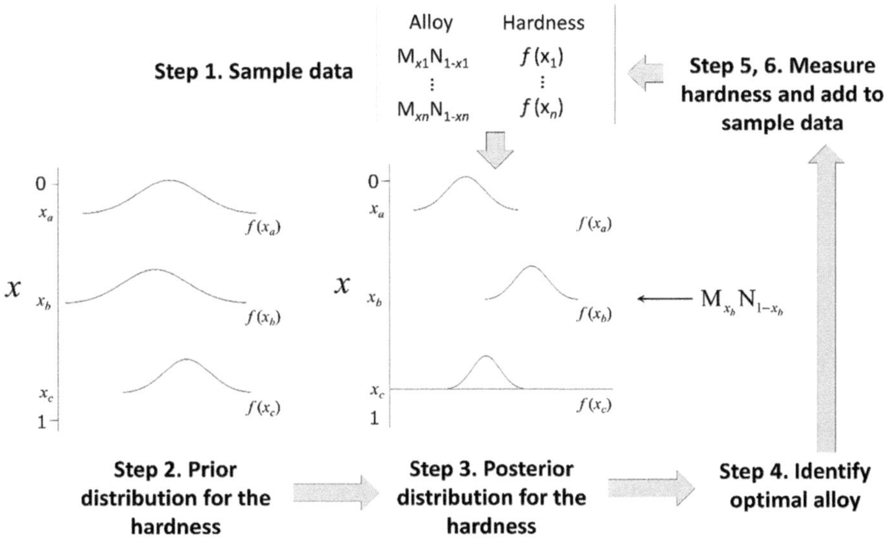

Fig. 1.2 Sketch of the Bayesian optimization method for finding the optimal stoichiometric coefficient x^*, for alloys of composition M_xN_{1-x}. Only a single iteration of the Bayesian optimization procedure is shown. The prior and posterior distributions are sketched for only three values of the stoichiometric coefficient (x_a, x_b, and x_c). See main text for details

Step 5. Let x_{n+1} be the value of x identified from Step 4. Create the alloy with stoichiometric coefficient x_{n+1} and measure the hardness $f(x_{n+1})$ of this new alloy.
Step 6. Add the new data x_{n+1} and $f(x_{n+1})$ to the sample data in Step 1.

Steps 2–6 are then repeated until the convergence to the optimal stoichiometry x^* is achieved. Bayesian optimization can be also applied in the same way to identify the stoichiometry which minimizes the hardness. It is possible to intuitively understand how Bayesian optimization works without going into technical details. The prior probability distribution in Step 2 measures our intuitive belief that an alloy with stoichiometric coefficient x has the hardness value $f(x)$. This prior distribution is built according to our prior expertise on M_xN_{1-x} alloys. Step 3 then 'corrects' the prior probability distribution by accounting for the information in the sample data.

The remarkable thing about Bayesian optimization is that it often able to identify the optimal value x^* within only small number of iterations of Steps 2–6. The origin of this excellent performance is in the fact that Bayesian optimization makes use of all information in the sample. In particular, the stoichiometric coefficients $x_1, x_2, \ldots, x_n$ in the sample data may be widely scattered between 0 and 1, which means that *global information* on the stoichiometric coefficient and hardness is being utilized. The algorithm intelligently jumps between different values of x until the global maximum x^* is identified, and is not so prone to converging in local minima. On the other hand, classical optimization algorithms scan the stoichiometric coefficients in such a way that the local derivative of hardness with respect to x approaches zero, and will inevitably converge in the nearest minimum that they can find.

While Bayesian optimization can potentially speed-up the materials fabrication process, it should not be treated as a black box. In fact, Bayesian optimization often performs poorly when the prior distribution is chosen without due consideration of the specific system at hand. On the other hand, Bayesian optimization is extremely difficult to perform unless a relatively simple prior distribution is chosen. In order to incorporate materials science expertize into Bayesian optimization, while maintaining mathematical tractability, we usually choose a normal distribution with mean $\mu\ (x)$ and standard deviation $\sigma\ (x)$ as the prior distribution. Here, $\mu\ (x)$ is an 'initial guess' for the hardness an alloy with stoichiometric coefficient x, and $\sigma\ (x)$ measures our uncertainty in this guess. For the case of a normal distribution for the prior distribution, the posterior distribution can be straightforwardly computed using the basic properties of Gaussian functions (see Chap. 2).

Note that Bayesian optimization using a normal distribution as the prior distribution is very closely related to the statistical techniques called Gaussian regression and Kriging. In fact, Gaussian regression is simply Steps 1–3 in the scheme described above.

1.2 Examples of Bayesian Optimisation in Materials Science

Here, we briefly summarise several studies which have applied Bayesian optimization to problems in materials science. Examples of Bayesian optimization in materials science have only appeared in the literature over the last couple of years, and so the literature on this subject is currently small.

1.2.1 Prediction of Compounds with Low Thermal Conductivity

The enormous gain in computational power over the last two decades has enabled *high-throughput screening* of large materials databases [2]. In these studies, several specific physical properties are calculated from first-principles (typically via density functional theory) for every material in the database, and the materials predicted to have the most desirable physical properties are then subject to experimental study. However, because these databases typically contain tens of thousands of candidate materials, high-throughput screening can only be performed with the physical properties of interest can be calculated within a short time period.

Lattice thermal conductivity (LTC) is an example of an important physical property which cannot be calculated to sufficient accuracy within such a short time period. Materials with low LTC are particularly desirable for a variety of applications. LTC results from anharmonic lattice dynamics and complex interactions between phonons. Consequently, an expensive combination of electronic structure theory and Boltzmann transport equation calculations are necessary to accurately predict the LTC. In order to discover new materials with low LTC via computational methods, a more efficient method than high-throughput screening must be employed.

In place of high-throughput screening, Seko et al. applied Bayesian optimization to discover new materials with low LTC [3]. In order to benchmark the performance of Bayesian optimization for these materials, the authors first considered a small library of 101 candidate materials selected from a crystal database. Due to the small size of this library, the LTC for each of the 101 materials could be calculated. Following this, the authors applied Bayesian optimization to this library, and showed that the material with the lowest LTC in this library could be identified within as few as 11 iterations of the Bayesian optimization procedure. Having benchmarked the performance of Bayesian optimization for these materials, the authors then applied Bayesian optimization to a much larger library containing 54799 candidate materials, and succeeded to predict 221 additional materials with very low LTC. Further filtering of these materials via additional first-principles calculations identified two materials (K_2CdPb and $Cs_2[PdCl_4]I_2$) with particularly suitable properties for device applications. The prediction of these materials would not be possible without an efficient method such as Bayesian optimization for scanning the enormous library of candidate materials.

1.2.2 Prediction of Compounds with Optimal Melting Temperatures and Elastic Properties

Continuing with the theme of high-throughput screening described above, Seko et al. applied Baysian optimization to predict materials with high melting temperatures [4]. From a library of 248 binary compounds of composition A_xB_y, where A and B are non-transition metal elements, 12 compounds were selected as an initial sample, and their melting temperatures were computed with DFT calculations. Based on this initial sample, Bayesian optimization was then applied to predict the compound with the highest melting temperature. The optimal compound from the library was identified within tens of repetitions of the Bayesian optimization procedure. In comparison, random sampling from the library required over 100 repetitions until the optimal material could be discovered. This result once again demonstrates the superior performance of Bayesian optimization in high-throughput screening studies.

Balachandran et al. applied a method similar Bayesian optimization to predict compounds with optimal elastic properties (i.e., very small and very large bulk, shear, and Young's moduli) [5]. They considered a library of 223 compounds of the form M_2AX, where M is a *d*-block element, and A and X are *p*-block elements. However, instead of creating a posterior distribution for the elastic properties of interest (as in Step 3 of the Bayesian optimization procedure), used a so-called *support-vector regression estimator* to fit the elastic properties in the sample data to the elastic properties of the materials. Using initial samples containing 20 materials, they could identify the material with the excellent elastic properties within tens of iterations. While this study did not use Bayesian optimization directly, it is often cited in the materials informatics literature and is similar in style to the work of Seko et al. described above.

1.2.3 Prediction of Interface Structures

A major challenge in computational materials science is the determination of the energy minimizing (ground state) atomic structure of complex materials. In the typical computational approach to this problem, structure relaxations (in which gradient-based energy minimizers are used to drive the atoms in the system into a local energy minimum) are routinely employed. This energy is usually calculated via DFT. However, the success of this approach depends upon the choice of initial atomic coordinates for the system. If the initial atomic structure does not lie in the region of the global energy minimum, then the structure relaxation will identify a metastable (local energy minimizing) atomic structure, but not the ground state structure. Unfortunately, in many materials the number of possible initial atomic structures is enormous, and often it is not possible to check every case, especially when the costs of each structure relaxation calculation may be quite high.

In addition to its use in virtual screening of material databases described in the previous examples, Bayesian optimization can be applied to structure optimizations as well. With Bayesian optimization, it is possible to quickly identify the 'optimal initial atomic structure' for a structure optimization, where 'optimal initial atomic structure' is the one which yields the ground state atomic structure following structural relaxation via DFT calculations. This method was employed by Kiyohara et al. in order to find energetically optimal grain boundary structures [6]. They considered the grain boundary formed by a Cu(001) phase and a Cu(210) phase, and generated 17,983 initial atomic structures by shifting the phases relative to each other via rigid-body translations. Bayesian optimization was then used to search through the initial atomic structures and identify the one which yields the minimum energy upon structure relaxation. Starting with a sample of 20 initial atomic structures and their energies upon structural relaxation, the authors could identify the most stable interface structure within as few as 49 iterations of the Bayesian optimization procedure (Fig. 1.3). This result is particularly remarkable considering that only around 0.4% of all possible initial atomic structures where examined.

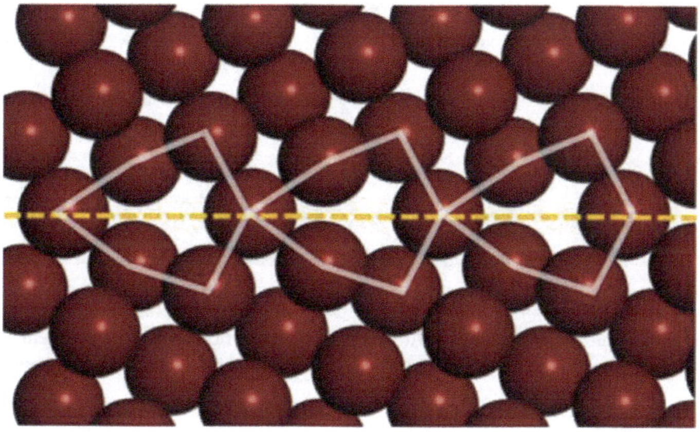

Fig. 1.3 Energy-minimising grain boundary structure predicted *via* Bayesian optimization by Kiyohara et al. [6]. The red spheres are Cu atoms, the dotted yellow line shows the position of the grain boundary, and the white polyhedra show the repeating structural unit of the grain boundary. Figure taken from [6]. Copyright 2017, The Japan Society of Applied Physics

1.2.4 Design of Interface Nanostructure

Another major challenge in materials science is to design materials with specific physical properties. In a typical problem of this type, we are given a small number of building blocks (atoms or parts of a molecule), and need to connect them together in such a way that we produce a material with the desired physical properties. Computational methods are very useful for studying these kinds of problems, as it is usually easy to generate a library of candidate materials by systematically connecting together the building blocks in silico. The material with the target physical property can then be identified by virtual screening of this library. However, as described above, virtual screening may not be viable approach if this library is particularly large.

In a study by Ju et al., Bayesian optimization was used to design nanostructures with high thermal conductances [7]. Specifically, they considered the interface formed by a Si crystal and a Ge crystal, and aimed to identify the atomic arrangement of Si and Ge atoms in the interface region which leads to the largest interfacial thermal conductance (ITC) (Fig. 1.4). By application of the open source Python library COMBO (=COMmon Bayesian Optimization) developed by Ueno et al. [8], the authors could identify interfacial structures with excellent ITCs by performing computations for around 438 candidate structures from a library of around 12,870 candidate structures. This is another remarkable result, considering that only around 3.4% of the entire library needed to be screened. A noteworthy part of this study is that the authors went beyond merely demonstrating power of Bayesian optimization for designing nanomaterials. In fact, from the interfacial structures predicted from their study, the authors could deduce new physical

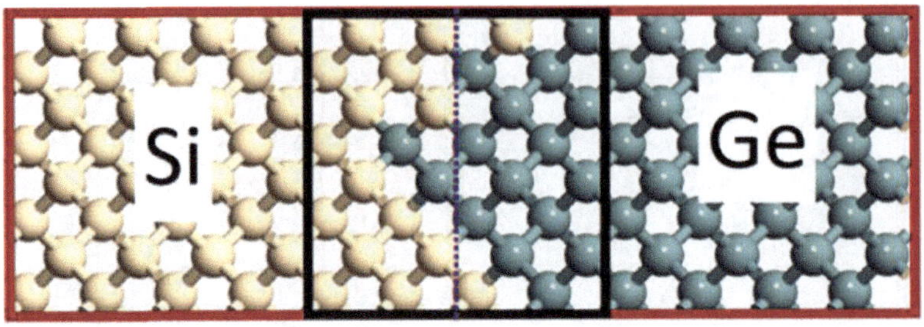

Fig. 1.4 Atomic structure of a nanostructure with high interfacial thermal conductance, as predicted via Bayesian optimization by Ju et al. [7]. The yellow spheres represent Si atoms, the green spheres represent Ge atoms, and the black box indicates the interfacial region. Figure from Ref. [7]

insights into thermal transport through the interfacial region, which in turn are expected to assist the development of other heat transporting materials in the future.

1.3 Bayesian Optimization Requires Good Feature Vectors

The above examples demonstrate that Bayesian optimization can indeed be an effective method for optimizing materials or material structures with respect to a specific property. However, how can one tell whether Bayesian optimization can be applied to a particular situation of interest? To discuss this point, let us consider the general situation in materials science, in which we have a large set of candidate materials, denoted as A, B, …, and Z, and we wish to identify the 'optimal' of these materials via Bayesian optimization.

In any application of Bayesian optimization, it is essential that each material can be described by a real-valued vector. These vectors are referred to as *feature vectors*. Let

$$\mathbf{r}(\mathrm{X}) = (r_1(\mathrm{X}), r_2(\mathrm{X}), \ldots, r_n(\mathrm{X})) \tag{1.1}$$

be the feature vector for material X. The components $r_1(\mathrm{X})$, $r_2(\mathrm{X})$, …, $r_n(\mathrm{X})$ are called *features* or *descriptors*, and these encode some physical information about material X. For example, if A, B, …, and Z are n-component alloys, then $r_k(\mathrm{X})$ might measure the fraction of component k in the alloy X. If instead A, B, …, Z each represent a possible unit cell structure for a specific compound, then $r_k(\mathrm{X})$ might be the coordinate of atom k in the unit cell. Bayesian optimization can only begin once we have encoded our candidate materials with an appropriate set of feature vectors.

Choosing appropriate feature vectors is the most important and often the most challenging part of Bayesian optimization, and considerable research activity is devoted to the development of feature vectors [9–11]. To understand the importance of choosing good feature vectors, suppose that our set of candidate materials A, B, …, and Z represent alloys, and we wish to identify the alloy with the highest electrical conductivity. If we apply Bayesian optimization to this problem, then we are making the implicit assumption that the conductivity is a function of the specific features chosen, and not of any other features. In other words, we must assume that

$$c_{\mathrm{X}} = f(r_1(\mathrm{X}), r_2(\mathrm{X}), \ldots, r_n(\mathrm{X})), \tag{1.2}$$

where c_{X} is the conductivity of candidate material X, f is some (unknown) function, and $r_k(\mathrm{X})$ is the kth feature of the feature vector chosen to represent X. The assumption in Eq. (1.2) will only be reliable if the features $r_1(\mathrm{X})$, $r_2(\mathrm{X})$, …, $r_n(\mathrm{X})$ have some meaningful relationship with the conductivity of X. If we choose features which are irrelevant to the conductivity of X, or if we fail to include some features which are important to the conductivity, then the assumption in Eq. (1.2) will become unreliable and Bayesian optimization will struggle to identify which material has the highest electrical conductivity.

A tempting way to get around the problem of choosing good features it to create very high dimensional feature vectors. In other words, we might try to create feature vectors by simply combining every conceivable property and variable that can be measured for the candidate materials. Unfortunately, Bayesian optimization using high dimensional feature vectors often performs very poorly. Continuing with the above example, suppose we choose to represent the alloys with feature vectors of the form

$$\mathbf{r}(\mathrm{X}) = (r_1(\mathrm{X}), r_2(\mathrm{X}), \ldots, r_n(\mathrm{X}), r_{n+1}(\mathrm{X}), r_{n+2}(\mathrm{X}), \ldots, r_{n+m}(\mathrm{X})). \tag{1.3}$$

Here, $r_1(\mathrm{X})$, $r_2(\mathrm{X})$, …, $r_n(\mathrm{X})$ represent the features which are important for determining the electrical conductivity of alloy X, and $r_{n+1}(\mathrm{X})$, $r_{n+2}(\mathrm{X})$, …, $r_{n+m}(\mathrm{X})$ represent the features which are not related to electrical conductivity. As we will see in the following chapter, the *distances* between feature vectors for different candidate materials play a critical role during the calculation of the posterior distribution. If we consider two candidate materials X and Y with very similar electrical conductivities, then the (squared) distance between their feature vectors can be written as

$$\|\mathbf{r}(\mathrm{X}) - \mathbf{r}(\mathrm{Y})\|^2 = \sum_{k=1}^{n} (r_k(\mathrm{X}) - r_k(\mathrm{Y}))^2 + \sum_{k=n+1}^{n+m} (r_k(\mathrm{X}) - r_k(\mathrm{Y}))^2. \tag{1.4}$$

The presence of the second term (which arises from the unimportant features) on the right-hand side of Eq. (1.4) increases the distance between $\mathbf{r}(\mathrm{X})$ and $\mathbf{r}(\mathrm{Y})$. Thus, even if the first term (which arises from the important features) is very small, the posterior distribution may incorrectly predict that X and Y have very different electrical conductivities. In turn, this will increase the number of iterations needed for Bayesian optimization to identify the optimal material. This problem might be not so severe in the case where the second term on the right-hand side of (1.4) does not vary so much between pairs of materials, however this situation is obviously ideal. In summary, in order to perform Bayesian optimization with sufficient efficiency, one should first construct low-dimensional feature vectors by identification of a minimum set of relevant features for the material property of interest.

The identification of a minimal set of features requires a deep understanding of materials science and experience with the particular type of system under study. This is a key point: Bayesian optimization is only a tool for aiding the materials development process, and does not eliminate the importance of genuine expertize in materials science.

References

1. Quote by President Obama, June 2011 at Carnegie Mellon University. See https://www.obamawhitehouse.archives.gov/mgi for more details.
2. Hinuma Y, et al. Discovery of earth-abundant nitride semiconductors by computational screening and high-pressure synthesis. Nat Commun. 2016;7:11962.
3. Seko A, et al. Prediction of low-thermal-conductivity compounds with first-principles anharmonic lattice-dynamics calculations and Bayesian optimization. Phys Rev Lett. 2015;115:205901.
4. Seko A, et al. Machine learning with systematic density-functional theory calculations: application to melting temperatures of single- and binary-component solids. Phys Rev B. 2014;89:054303.
5. Balachandran PV, et al. Adaptive strategies for materials design using uncertainties. Sci Rep. 2016;6:19660.
6. Kiyohara S, et al. Acceleration of stable interface structure searching using a Kriging approach. Jpn J Appl Phys. 2016;55:045502.
7. Ju S, et al. Designing nanostructures for photon transport *via* Bayesian optimization. Phys Rev X. 2017;7:021024.
8. Ueno T, et al. COMBO: An efficient Bayesian optimization library for materials science. Mater Discov. 2016;4:18.
9. Rupp M, et al. Fast and accurate modeling of molecular atomization energies with machine learning. Phys Rev Lett. 2012;108:058301.
10. Hansen K, et al. Assessment and validation of machine learning methods for predicting molecular atomization energies. J Chem Theory Comput. 2013;9:3404.
11. Huo, H, Rupp, M. Unified representation for machine learning of molecules and crystals. arXiv:1704.06439v1 [physics.chem-ph].

Chapter 2
Theory of Bayesian Optimization

In this chapter, we introduce the theory of Bayesian optimization procedure and illustrate its application to a simple problem. A more involved application of Bayesian optimization will be presented in Chap. 3.

2.1 Bayesian Interpretation of Probability

Consider rolling a die with k sides, labeled a_1, a_2, …, and a_k, respectively. Let $P(a_j)$ be the 'probability' that a particular side a_j appears after rolling the die. Before attempting to calculate $P(a_j)$, it is necessary to clarify the meaning of the word 'probability'. In other words, we must specify what the number $P(a_j)$ quantifies. In statistics, the concept of 'probability' is formally interpreted in one of two ways. In the *frequentist interpretation of probability*, $P(a_j)$ is the fraction of times that a_j appears in a very large number of die rolls. In the *Bayesian interpretation of probability*, $P(a_j)$ is the extent to which we believe that the number a_j will appear prior to rolling the die.

For the case of a die with k sides, there is little difference between the frequentist and Bayesian interpretation of probability. Given that the die is not biased in any way, we would set $P(a_j) = 1/k$ in both the frequentist and Bayesian interpretations. However, a major difference between the frequentist and Bayesian interpretation of probability arises when we consider so-called *learning-type problems*, in which new information on the system becomes available over time. For example, consider a robot whose job is to sort oranges from lemons. Suppose that the robot is presented with a fruit, and that the robot has no useful information to help distinguish between oranges and lemons. For example, the robot does not know that round fruit are more likely to be oranges rather than lemons. In this case, the robot would set $P(\text{o}) = 1/2$ and $P(\text{l}) = 1/2$, where $P(\text{o})$ and $P(\text{l})$ are the probabilities that the fruit is an orange or lemon, respectively. Now, suppose that new information is loaded into the robot's memory from an external source, namely

D. Packwood, *Bayesian Optimization for Materials Science*,
SpringerBriefs in the Mathematics of Materials 3,
https://doi.org/10.1007/978-981-10-6781-5_2

$$L(\mathrm{r}|\mathrm{o}) = 8, \tag{2.1}$$

and

$$L(\mathrm{r}|\mathrm{l}) = 1. \tag{2.2}$$

These numbers are called *likelihoods*, and result from measurements on different types of fruits by the external source. $L(\mathrm{r}|\mathrm{o})$ measures the 'likelihood' that an orange is round, and $L(\mathrm{r}|\mathrm{l})$ measures the 'likelihood' that a lemon is round. The precise physical meaning of 'likelihood' and it units are do not need to made so clear, providing that the values of the likelihoods are always measured in a consistent way. In order to utilize the information provided by the likelihoods, we employ a formula called *Bayes' rule*. Bayes' rule can be written as

$$P(\mathrm{o}|\mathrm{r}) \propto L(\mathrm{r}|\mathrm{o})P(\mathrm{o}) \tag{2.3}$$

$$P(\mathrm{l}|\mathrm{r}) \propto L(\mathrm{r}|\mathrm{l})P(\mathrm{l}), \tag{2.4}$$

where $P(\mathrm{o}|\mathrm{r})$ and $P(\mathrm{l}|\mathrm{r})$ are the probability that a round fruit is an orange, and the probability that a round fruit is a lemon, respectively. Substituting the numbers given above, we find that $P(\mathrm{o}|\mathrm{r}) \propto 8 \times 0.5 = 4$ and $P(\mathrm{l}|\mathrm{r}) \propto 1 \times 0.5 = 0.5$. Eliminating the proportionality constants then gives $P(\mathrm{o} \mid \mathrm{r}) = 4/(4 + 0.5) = 0.89$ and $P(\mathrm{o}|\mathrm{r}) = 0.5 / (4 + 0.5) = 0.11$. Thus, when presented with a round fruit, the robot will determine that there is a probability of 0.89 that the fruit is an orange and a 0.11 probability that the fruit is a lemon. With Bayes' rule, the robot is therefore able to improve its ability to classify fruit by incorporating information provided by an external source. This kind of process is not natural within the frequentist interpretation of probability, in which the probabilities $P(\mathrm{o})$, $P(\mathrm{r})$, $P(\mathrm{o}|\mathrm{r})$ and $P(\mathrm{l}|\mathrm{r})$ remain fixed for all time, regardless of any new information which may appear.

Within the Bayesian interpretation of probability, $P(\mathrm{o})$ and $P(\mathrm{l})$ are referred to as *prior probabilities* and $P(\mathrm{o}|\mathrm{r})$ and $P(\mathrm{l}|\mathrm{r})$ are referred to as *posterior probabilities*. Note that the likelihoods $L(\mathrm{r}|\mathrm{o})$ and $L(\mathrm{r}|\mathrm{l})$ in Eqs. (2.1) and (2.2) can be regarded as a function of the type of fruit. For this reason, L is referred to as a *likelihood function*.

As one might have guessed, Bayesian optimization makes use of the Bayesian interpretation of probability and Bayes' rule. We elaborate upon this point further in the following section.

2.2 Equilibrium Bond Lengths *Via* Bayesian Optimization

Consider the problem of estimating the equilibrium bond length r_0 of a diatomic molecule such as Br_2. By 'equilibrium bond length', we mean that the interatomic potential energy $u(r)$ is minimized when $r = r_0$. We suppose that the analytical form of $u(r)$ is unknown, i.e., that we cannot find r_0 by directly differentiating a simple

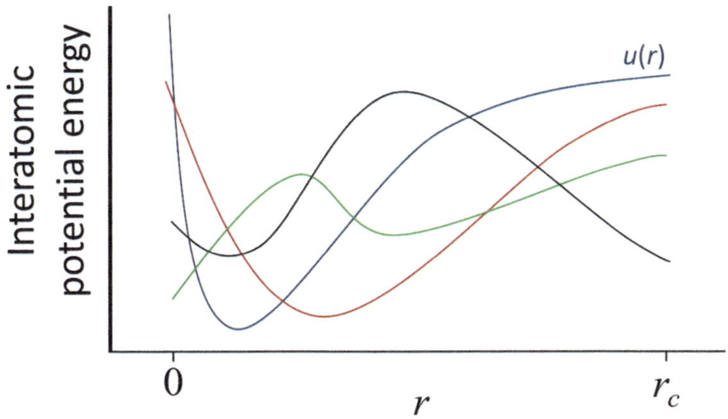

Fig. 2.1 Sketch of the sample space Ω (see main text). Only four candidate functions are shown. The function shown in blue corresponds to the true interatomic potential for the diatomic molecule. The sample space contains all candidate functions that are real-valued and differentiable between 0 and r_c. This includes functions which are physically unreasonable, such as the ones shown in green and black

formula. While the analytical form of $u(r)$ is unknown, it is a physical requirement that $u(r)$ be real-valued, continuous, and differentiable over the interval $0 \leq r \leq r_c$, where, r_c is the dissociation limit of the molecule. For simplicity, we assume that r_c is a well-defined and known constant. We define the *sample space* Ω as the collection of all real-valued functions which continuous and differentiable over the interval $0 \leq r \leq r_c$ (Fig. 2.1). In the present context, Ω can be thought of as a collection of 'candidate functions' for the interatomic potential, one of which corresponds to the true interatomic potential, u. Ω contains an infinite number of functions.

Estimation of the equilibrium bond length r_0 via Bayesian optimization runs according to the following steps. (i) Generate a random sample of interatomic separations and measure (or calculate, from first-principles) u for each case. (ii) Independently of (i), assign a prior probability to the functions in Ω. The prior probability measures our intuitive feeling about which functions in Ω correspond to the true interatomic potential. (iii) Use the sample data and Bayes' rule to calculate the posterior probability for the functions in Ω. (iv) Use the posterior distribution to estimate r_0. The estimated value of r_0 is denoted r^*. (*v*) Measure (or calculate from first-principles) $u(r^*)$, the interatomic potential at distance r^*, and add r^* and $u(r^*)$ to the sample data. (vi) Repeat steps (ii)–(v) until the global minimum of the interatomic potential is identified (i.e., when the minimum value of u in the sample remains unchanged over several iterations).

Note that, strictly speaking, Bayesian optimization assumes that all functions in the sample space are finite. This assumption is actually violated for the present system, because it is a physical requirement that $u(0) = \infty$ for the true interatomic potential u. In the present analysis, we will get around this issue by simply supposing that $u(0)$ is finite.

2.2.1 Prior Probability

In Bayesian optimization, we choose a multivariate Gaussian distribution for the prior probability distribution. In other words, choose s interatomic separations r_i, r_j, …, r_k (where $0 \leq r_l \leq r_d$ for $l = i, j, \ldots,$ or k). Let $u(r_i)$, $u(r_j)$, …, and $u(r_k)$ be the values of the true interatomic potential at points r_i, r_j, …, and r_k. The prior probability that the vector $(u(r_i), u(r_j), \ldots, u(r_k))$ is contained in an infinitesimal region of space centered at point $\mathbf{v} = (v_i, v_j, \ldots, v_j)$ is given by $g(v_i, v_j, \ldots, v_k)dv_i dv_j \cdots dv_k$, where

$$g(v_i, v_j, \ldots, v_k) = \frac{1}{(2\pi)^{s/2}|\mathbf{K}|^{1/2}} \exp\left(-\frac{1}{2}(\mathbf{v} - \boldsymbol{\mu})^T \mathbf{K}^{-1}(\mathbf{v} - \boldsymbol{\mu})\right) \tag{2.5}$$

is referred to as the *prior probability density*. In Eq. (2.5), the $s \times 1$ column matrix $\boldsymbol{\mu}$ is called the *mean vector*, the $s \times s$ matrix $\mathbf{K}$ is called the *covariance matrix*, and $|\mathbf{A}|$ denotes the determinant of matrix $\mathbf{A}$. In Eq. (2.5), $\mathbf{v}$ is treated as an $s \times 1$ column vector. If $g(v_i, v_j, \ldots, v_k)$ is particularly large, then it means that we have a strong intuitive feeling that $u(r_i) = v_i$, $u(r_j) = v_j$, …, and $u(r_k) = v_k$ for the true interatomic potential energy function u.

Our intuitive beliefs about the interatomic potential u are encoded into the prior distribution through the mean vector $\boldsymbol{\mu}$ and covariance matrix $\mathbf{K}$. The choice of $\boldsymbol{\mu}$ and (particularly) $\mathbf{K}$ has a large influence on the performance of Bayesian optimization, and therefore they should be carefully considered before applying Bayesian optimization to a physical problem.

If we write $\boldsymbol{\mu} = (\mu_i, \mu_j, \ldots, \mu_k)$ for the mean vector, then μ_i can be regarded as our intuitive guess for $u(r_i)$, the actual value of the interatomic potential at interatomic separation r_i. For example, we might choose the harmonic oscillator potential as an initial guess for u, and write

$$\mu_i = \frac{1}{2}C(r_i - R_0)^2, \tag{2.6}$$

where the parameters C and R_0 are guessed by considering literature data for similar diatomic molecules. There are no particular restrictions on the values of μ_i, however they must be finite.

Let us write $\mathbf{K} = [K_{ij}]_{s \times s}$ for the covariance matrix. If V is chosen at random from Ω according to the prior distribution, then K_{ij} measures the extent to which we believe $V(r_i)$ should be correlated with $V(r_j)$. To quantify this correlation, let L be an intuitive guess for the correlation length of the interatomic potential $u(r)$. Roughly speaking, L measures the length over which r must change in order for $u(r)$ to change significantly. Returning now to K_{ij}, we would expect for K_{ij} to be large when $|r_i - r_j| < L$, and moreover K_{ij} should decrease rapidly as $|r_i - r_j|$ increases beyond L. This behavior can be acquired by choosing a squared exponential function for K_{ij}, i.e.,

$$K_{ij} = a\exp\left(-\frac{\left|r_i - r_j\right|^2}{2L^2}\right), \tag{2.7}$$

where a is another constant. If a single point $V(r_i)$ is randomly generated from Ω in accordance with the prior distribution, then a is interpreted as the mean-square deviation of $V(r_i)$ from μ_i. This interpretation follows from the fact that the diagonal elements of **K** formally correspond to the variance of $V(r_i)$, when V is randomly generated from the prior distribution.

The constants a and L are referred to as *hyperparameters* and have a critical influence on the performance of Bayesian optimization. While a and L can be also chosen based on our intuitive feelings about the system, in practice it is preferable to choose them via a training procedure. We will describe this in Sect. 2.6 and in the following chapter.

Note that Bayesian optimization is not restricted to the covariance matrix defined in Eq. (2.7). Mathematically speaking, it is only necessary for the covariance matrix to be positive semidefinite. Some alternative forms of the covariance matrix are discussed in reference [1]. The advantage of Eq. (2.7) is that it physical interpretation is relatively straightforward.

2.2.2 *Likelihood Function and Posterior Distribution*

In Bayesian optimization, the likelihood function is assumed to be the same Gaussian density function as was used for the prior probability density in Eq. (2.5). To explain what is meant here, let us start by re-writing Eq. (2.5) as

$$\begin{aligned} &g\left(v_\alpha, v_\beta, \ldots, v_\gamma, v_i, v_j, \ldots, v_k\right) \\ &= \frac{1}{(2\pi)^{\frac{m+s}{2}}|\mathbf{K}|^{1/2}}\exp\left(-\frac{1}{2}\left[\begin{pmatrix}\mathbf{v}_{\alpha:\gamma}\\ \mathbf{v}_{i:k}\end{pmatrix} - \begin{pmatrix}\boldsymbol{\mu}_{\alpha:\gamma}\\ \boldsymbol{\mu}_{i:k}\end{pmatrix}\right]^T \begin{bmatrix}\mathbf{K}_{\alpha:\gamma,\alpha:\gamma} & \mathbf{K}_{\alpha:\gamma,i:k}\\ \mathbf{K}_{i:k,\alpha:\gamma} & \mathbf{K}_{i:k,i:k}\end{bmatrix}^{-1}\left[\begin{pmatrix}\mathbf{v}_{\alpha:\gamma}\\ \mathbf{v}_{i:k}\end{pmatrix} - \begin{pmatrix}\boldsymbol{\mu}_{\alpha:\gamma}\\ \boldsymbol{\mu}_{i:k}\end{pmatrix}\right]\right) \end{aligned} \tag{2.8}$$

In Eq. (2.8), $\mathbf{v}_{\alpha:\gamma} = (v_\alpha, v_\beta, \ldots, v_\gamma)$, $\mathbf{v}_{i:k} = (v_i, v_j, \ldots, v_k)$, m is the length of the vector $\mathbf{v}_{\alpha:\gamma}$, s is the length of the vector $\mathbf{v}_{i:k}$, $\mathbf{K}_{\alpha:\gamma,\alpha:\gamma}$ is the covariance matrix for points r_α, r_β, …, and r_γ, $\mathbf{K}_{i:k,i:k}$ as the covariance matrix for points r_i, r_j, …, and r_k, and

$$\mathbf{K}_{\alpha:\gamma,i:k} = \begin{bmatrix} K_{\alpha i} & K_{\alpha j} & \cdots & K_{\alpha k}\\ K_{\beta i} & K_{\beta j} & \cdots & K_{\beta k}\\ \vdots & \vdots & \ddots & \vdots\\ K_{\gamma i} & K_{\gamma j} & \cdots & K_{\gamma k} \end{bmatrix} \tag{2.9}$$

Note that $\mathbf{K}_{i:k,\alpha:\gamma}$ is the transpose of $\mathbf{K}_{\alpha:\gamma,i:k}$. The likelihood function used in Bayesian optimization is defined as

$$L(v_i, v_j, \ldots, v_k | v_\alpha, v_\beta, \ldots, v_\gamma) = g(v_i, v_j, \ldots, v_k | v_\alpha, v_\beta, \ldots, v_\gamma). \tag{2.10}$$

Here, $g(v_i, v_j, \ldots, v_k | v_a, v_\beta, \ldots, v_\gamma)$ is a so-called *conditional density*. It corresponds to the prior probability density in Eq. (2.8) calculated a point $(v_i, v_j, \ldots, v_k)$ with the values of v_α, v_β, …, and v_γ held fixed. An analytic formula for the conditional density can be written, however it turns out that this formula is not necessary for our purposes (see Appendix 2.1).

Let us now suppose that we are provided with a sample of s points $(r_i, u(r_i))$, $(r_j, u(r_j))$, …, $(r_k, u(r_k))$, where for $t = i, j, \ldots$, and k, we have $0 < r_t < r_d$ and the value of $u(r_t)$ is known exactly. In analogy to Eqs. (2.3) and (2.4), the posterior probability that the vector $(u(r_\alpha), u(r_\beta), \ldots, u(r_\gamma))$ is contained in an infinitesimal region of space centered at point $\mathbf{v} = (v_\alpha, v_\beta, \ldots, v_\gamma)$ is given by Bayes' rule, namely

$$\begin{aligned} &f(v_\alpha, v_\beta, \ldots, v_\gamma | u_i, u_j, \ldots, u_k) dv_\alpha dv_\beta \cdots dv_\gamma \\ &\quad \propto L(u_i, u_j, \ldots, u_k | v_\alpha, v_\beta, \ldots, v_\gamma) g(v_\alpha, v_\beta, \ldots, v_\gamma) dv_\alpha dv_\beta \cdots dv_\gamma, \end{aligned} \tag{2.11}$$

where we have used the notation $u_i = u(r_i)$. $f(v_\alpha, v_\beta, \ldots, v_\gamma | u_i, u_j, \ldots, u_k)$ is referred to as the *posterior probability density*. Substituting Eqs. (2.5) and (2.10) into Eq. (2.11) and performing various manipulations, we obtain (see Appendix 2.1),

$$f(v_\alpha, v_\beta, \ldots, v_\gamma | u_i, u_j, \ldots, u_k) \propto \exp\left(-\frac{1}{2}\left[\mathbf{v}_{\alpha:\gamma} - \boldsymbol{\mu}^*_{\alpha:\gamma}\right]^T \left(\mathbf{K}^*_{\alpha:\gamma,\alpha:\gamma}\right)^{-1} \left[\mathbf{v}_{\alpha:\gamma} - \boldsymbol{\mu}^*_{\alpha:\gamma}\right]\right), \tag{2.12}$$

where

$$\boldsymbol{\mu}^*_{\alpha:\gamma} = \boldsymbol{\mu}_{\alpha:\gamma} - \mathbf{K}_{\alpha:\gamma,i:k}\mathbf{K}^{-1}_{i:k,i:k}(\mathbf{u}_{i:k} - \boldsymbol{\mu}_{i:k}), \tag{2.13}$$

and

$$\mathbf{K}^*_{\alpha:\gamma,\alpha:\gamma} = \mathbf{K}_{\alpha:\gamma,\alpha:\gamma} - \mathbf{K}_{\alpha:\gamma,i:k}\mathbf{K}^{-1}_{i:k,i:k}\mathbf{K}_{i:k,\alpha:\gamma}. \tag{2.14}$$

The right-hand side of Eq. (2.12) is actually the unnormalized multivariate Gaussian distribution. If we are interested in the posterior density a single point v_α, then Eq. (2.12) simplifies to

$$f(v_\alpha | u_i, u_j, \ldots, u_k) = \frac{1}{\sqrt{2\pi K^*_{\alpha\alpha}}} \exp\left(-\frac{(v_\alpha - \mu^*_\alpha)^2}{2K^*_{\alpha\alpha}}\right), \tag{2.15}$$

where

$$\mu_{\alpha}^{*} = \mu_{\alpha} - \mathbf{K}_{\alpha,i:k}\mathbf{K}_{i:k,i:k}^{-1}(\mathbf{u}_{i:k} - \boldsymbol{\mu}_{i:k}), \tag{2.16}$$

$$K_{\alpha\alpha}^{*} = K_{\alpha\alpha} - \mathbf{K}_{\alpha,i:k}\mathbf{K}_{i:k,i:k}^{-1}\mathbf{K}_{i:k,\alpha}, \tag{2.17}$$

the row-vector $\mathbf{K}_{\alpha,i:k}$ is defined as $(K_{\alpha,i}, K_{\alpha,j}, \ldots, K_{\alpha,k})$, and $\mathbf{K}_{i:k,\alpha}$ is the transpose of $\mathbf{K}_{\alpha,i:k}$. In practice, we use Eqs. (2.15–2.17) in all calculations of the posterior distribution.

2.2.3 Example Calculation of the Posterior Distribution

Figure 2.2a–c plot the mean and variance of the posterior distribution from Eqs. (2.16) and (2.17) calculated from a sample of three interatomic displacements for an isolated Br_2 molecule. The red line represents the posterior mean, the thin black lines measure the posterior variance and correspond to the 95% confidence limits of the posterior distribution, i.e.,

$$\mu_{\alpha}^{*} \pm 1.96\sqrt{\mathbf{K}_{\alpha,\alpha}^{*}},$$

and the blue line shows the actual interatomic potential energy curve. The potential energy at each point was calculated from first principles using density functional theory (DFT). DFT calculations discussed here and elsewhere in this chapter were performed with the VASP code [2], using a plane wave basis set, projector-augmented wave (PAW) potentials, and the generalized gradient approximation (GGA).

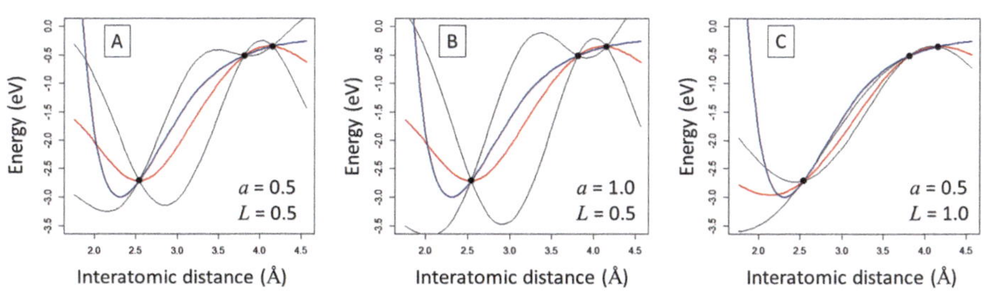

Fig. 2.2 Calculation of the posterior distribution for the interatomic potential energy for a Br_2 molecule, using a sample of three interatomic distances and the corresponding energies and various values of the hyperparameters a and L. The blue curve corresponds to the true interatomic potential energy, the red curve correspond to the mean of the posterior distribution (Eq. 2.16), and the black curves correspond to the 95% confidence limits of the posterior distribution. See Sect. 2.3 for details. Note that the y axis actually plots the total energy of the Br_2 molecule, which is equal to the interatomic potential energy plus a constant

By looking at the sample points in Fig. 2.2a–c, it is clear that the posterior mean interpolates the sample points exactly, and that the posterior variance is zero at the sample points. This is a general behavior of Bayesian optimization, and shows that the posterior distribution predicts the sample data exactly.

An important observation is that both the posterior mean and variance depend on the values of the hyperparameters a and L in the covariance matrix in Eq. (2.7). In general, the posterior variance grows and then shrinks back to zero as we move between successive sample points. As the hyperparameter L increases and the 'correlation length' of the system grows, the posterior variance grows more slowly between successive sample points and the posterior mean begins to resemble a linear interpolation between successive sample points. In this sense, when the correlation length in the system is assumed to be large, we become more confident that true potential energy curve can be obtained by a linear interpolation between successive points. In Fig. 2.2c, in which the correlation length is large, it can be seen that the true potential energy curve (blue) actually lies outside of the 95% confidence limits of the Gaussian distribution, showing that Bayesian optimization predicts a very small probability for the true interatomic potential when the correlation length L is large. In general, we should choose the hyperparameters so that the true interatomic potential lies within the 95% confidence limits of the Gaussian distribution. In this situation, the posterior density will be large for functions closely resembling the true interatomic potential, and the Bayesian optimization procedure will be able to accurately estimate the location of the minimum of the true potential curve.

2.2.4 The Expected Improvement

Having generated the posterior distribution from the sample data, we now need to predict the point which minimizes the interatomic potential energy. There are a variety of ways of predicting the position of the optimum using the posterior distribution. One of the most popular methods is involves the *expected improvement*, which we consider here.

The expected improvement at point r_α is defined as

$$EI(r_\alpha) = E_f\left[\max\left(u_{\min} - V(r_\alpha), 0\right)\right], \tag{2.18}$$

where $E_f[A]$ is the expected value (average) of the random variable A with respect to the posterior distribution in Eq. (2.18), and $u_{\min}$ is the minimum interatomic potential energy in the sample. $V(r_\alpha)$ is the value of a function V evaluated at point r_α, where V has been randomly generated from the sample space according to the posterior distribution. The interatomic distance which minimizes the interatomic potential energy is then estimated as the value of r_α which maximizes Eq. (2.18). Thus, the expected improvement considers our current 'best guess' of the minimum interatomic potential, $u_{\min}$, and then determines the point which, on average, will improve upon that guess the most.

In order to calculate the expected improvement, the following formula may be used (see Appendix 2.2),

$$EI(r_\alpha) = \left(u_{\min} - \mu_\alpha^*\right)\Phi\left(\frac{u_{\min} - \mu_\alpha^*}{\sqrt{K_{\alpha\alpha}^*}}\right) + \sqrt{K_{\alpha\alpha}^*}\varphi\left(\frac{u_{\min} - \mu_\alpha^*}{\sqrt{K_{\alpha\alpha}^*}}\right). \quad (2.19)$$

In Eq. (2.19), $\Phi(x)$ and $\varphi(x)$ are the normal distribution function and normal density function, respectively, evaluated at point x. Both functions can be can be easily called in a statistical programming environment such as *R* [3].

2.2.5 Example Run of Bayesian Optimisation

In order to demonstrate the calculation of the expected improvement, and to show the Bayesian optimization procedure in action, we return to the example of the Br_2 molecule discussed at the end of the previous section. Figure 2.3a plots the posterior mean (red line) and confidence limits (black lines) from a sample of two interatomic distances, using hyperparameter values $a = 0.5$ and $L = 0.5$ and energies calculated via DFT. The green line represents the expected value calculated from Eq. (2.18). The peak of the expected improvement lies at 2.25 Å. The true interatomic potential energy $u(r)$ is then calculated for the interatomic distance $r = 2.25$ Å via DFT, and this data is added to the sample. Figure 2.3b plots the posterior mean, confidence limits, and expected improvement for the new sample. This time, the expected improvement peaks at 2.27 Å, and so the true interatomic potential energy is calculated at this interatomic displacement, and this data is added to the sample. After repeating this procedure only a few more times (Fig. 2.3c, d), the expected improvement peaks at 2.30 Å (Fig. 2.3e). This corresponds to the exact optimum interatomic bond length for the Br_2 molecule (within the accuracy of the present DFT method), showing that the calculation has converged to the global optimum within relatively few iterations of the Bayesian optimization procedure.

A close look Fig. 2.3a–e unveils a key feature of Bayesian optimization. Comparing the expected improvement calculated at successive rounds of the Bayesian optimization procedure, we see that the added sample points are widely scattered and are not localized at any particular point. For example, at the end of the first, second, and third rounds of Bayesian optimization (Fig. 2.3a–c), the expected value peaks at 2.58, 2.27, and 1.76 Å, respectively, and these values and the corresponding interatomic potential energy are added to the sample data. This shows that Bayesian optimization is a non-local search method, which is in contrast with conventional optimizers which rely on a local gradient. The reason for the non-locality of Bayesian optimization is that the posterior distribution in Eq. (2.15) is computed by utilizing *all* information in the sample, which may be scattered

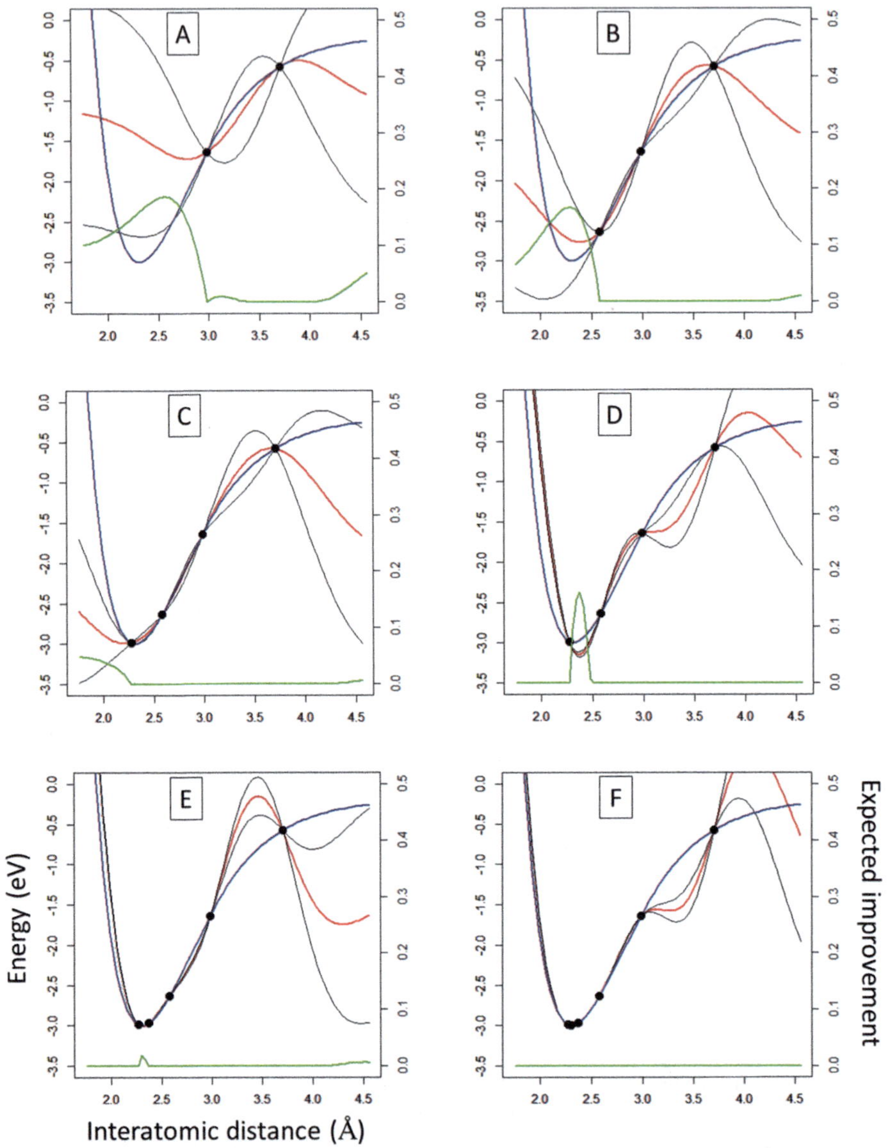

Fig. 2.3 Bayesian optimization procedure for estimating the interatomic displacement which minimizes the interatomic potential of a Br_2 molecule. Blue, red, black, and green curves correspond to the true potential energy curve, the mean of the posterior distribution, the 95% confidence limits of the posterior distribution, and the expected improvement, respectively. Starting with a sample of two interatomic displacements and their energies (**a**), sample data is successively added according to the maximum of the expected improvement. Note that the point added D is not shown, as it appears beyond the scale of these plots. Hyperparameter values of $a = 0.5$ and $L = 0.5$ were used. See main text for details

around the space. The non-locality of Bayesian optimization makes it less prone (but not immune) to getting trapped in local minima.

Figure 2.3a–e also demonstrate the so-called *exploitation versus exploration trade-off* concept [4]. The expected improvement tends to grow as the posterior mean decreases and the posterior variance increases. The former effect encourages investigation of 'promising' regions ('exploitation'), on the basis of information contained in the sample, whereas the latter effect encourages the exploration of regions in which we have little sample information ('exploration'). Exploitation is evident in Fig. 2.3f and e, in which the general region of the potential minimum becomes apparent and the search focuses on this region. Exploration is evident in Fig. 2.3c, in which the relatively small interatomic distance of 1.76 Å is suddenly added to the sample, causing the Bayeisan optimization procedure to gather information on the system at very small interatomic distance. The extent of exploration versus exploitation is determined by the posterior mean and variance, which in turn are strongly affected by the hyperparameters a and L. This shows once again the importance of the hyperparameters in determining the effectiveness of the Bayesian optimization procedure.

The performance of Bayesian optimization is further shown in Fig. 2.4. Figure 2.4a plots the minimum interatomic potential energy and optimal interatomic distance for the Br_2 molecule as a function of the sample size used in the calculation of the posterior distribution. The sample size corresponds to the number of DFT calculations. Because the initial sample contained 2 points, the number of iterations of Bayesian optimization is equal to the sample size −1. In Fig. 2.4b, the minimum interatomic potential energy and optimal interatomic distance for the case of random sampling from a grid of 83 interatomic displacements between 1.76 and 4.46 Å. For the latter calculations, the minimum interatomic potential energies and

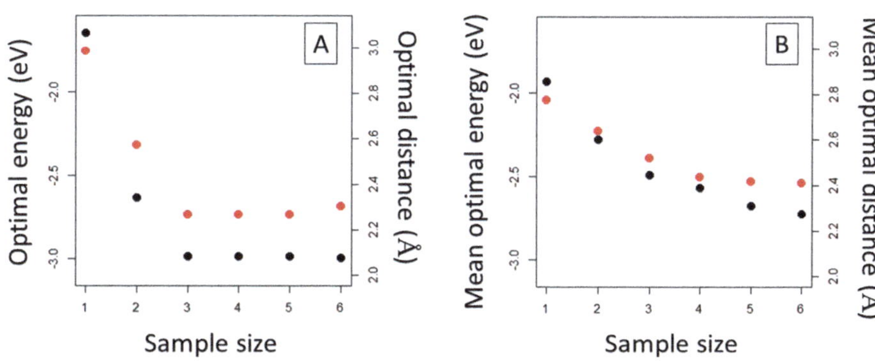

Fig. 2.4 **a** Minimum interatomic potential energy (black points) and corresponding interatomic distance (red points) predicted by the Bayesian optimization procedure in Fig. 2.3. The interatomic distance at the true minimum of the potential energy curve is 2.30 Å. **b** Minimum interatomic potential energy and corresponding interatomic distance predicted from random sampling interatomic displacements (see text for details). The data points in (b) have been averaged across 100 independent rounds of random sampling

optimal interatomic distances have been averaged across 100 rounds of sampling. While Bayesian optimization is able to find the optimal interatomic displacement (2.30 Å) within 5 iterations of the procedure, random sampling, on average, predicts a much larger value of around 2.42 Å for the same sample size.

2.2.6 *Training*

The discussion and the end of the previous section demonstrates that the performance of the Bayesian optimization procedure is heavily affected by the choice of parameters a and L for the covariance matrix. In the context of Bayesian optimization, *training* refers to a procedure for choosing the 'best' values of a and L for the prior distribution. Here, 'best' refers to the values of a and L which result in the quickest identification of the global optimum for a given sample of data. While the 'best' values of a and L can often be determined by physical intuition, it is common (although not necessarily more reliable) to choose these values by a more statistical approach. One of the most standard of these statistical approaches is referred to as *marginal likelihood maximization* (MLM), which we introduce here.

Continuing with the problem of finding the equilibrium bond distance in a diatomic molecule, suppose again that we have a sample of s interatomic bond distances and the corresponding energies, $(r_i, u(r_i))$, $(r_j, u(r_j))$, …, $(r_k, u(r_k))$. In the MLM approach, we find the values of the hyperparameters a and L which maximize the value of the prior distribution in Eq. (2.5) when calculated for the points in the sample data. More precisely, we wish to obtain the values of the hyperparameters which maximize $g(u_i, u_j,\ldots, u_k)$, where g is the prior distribution in Eq. (2.5) and $u_i = u(r_i)$, $u_j = u(r_j)$, …, and $u_k = u(r_k)$ are the sample data for the interaction potential.

For the special case of a constant prior mean $\boldsymbol{\mu}$ and a squared exponential function (as in Eq. (2.7)) for the covariance function, we can use the following equations to maximize the prior distribution, namely (see Appendix 2.3)

$$\log g(u_i, u_j, \ldots, u_k) = -\frac{s}{2}\log\left(\frac{1}{s}|\mathbf{R}|^{1/s}(\mathbf{u}-\boldsymbol{\mu})^T\mathbf{R}^{-1}(\mathbf{u}-\boldsymbol{\mu})\right) + c, \tag{2.20}$$

where c is a term which is independent of a and L, and

$$a = \frac{1}{s}(\mathbf{u}-\boldsymbol{\mu})^T\mathbf{R}^{-1}(\mathbf{u}-\boldsymbol{\mu}), \tag{2.21}$$

where $\mathbf{R} = [R_{ij}]_{s\times s}$ and

$$R_{ij} = \exp\left(-\frac{(r_i - r_j)^2}{2L^2}\right). \tag{2.22}$$

First, the value of L which maximizes the right-hand size of Eq. (2.20) is identified by numerically computing Eq. (2.20) across a grid of candidate values of L. This value of L is then substituted directly into Eq. (2.21) to obtain the optimum value of a.

For the special case of the Br_2 molecule studied here, the use of MLM approach to obtain the hyperparameters does not improve the performance of Bayesian optimization compared to the results discussed at the end of Sect. 2.5. However, the MLM approach can be useful for situations in which more difficult systems are studied and it is not possible to guess a value of a or a typical 'correlation distance' from intuition. In any case, because hyperparameters have a critical influence on the performance of Bayesian optimization, we strongly advocate for the use of a physically motivated procedure to estimate good values for a and L. Such a procedure is discussed in the following chapter.

2.3 Bayesian Optimization in the General Case

The above formalism can be immediately applied to systems beyond a simple diatomic molecule. In the general case, we have n *objects*, x_1, x_2, …, x_n, where object x_k has a *property* $h(x_k)$. We wish to identify the object whose property has the minimum value. These objects may be different types of materials or different configurations of molecules, and the properties may be material properties such as thermal conductivity or molecular properties such as HOMO energy level.

In order to apply the formalism above to the general case, we simply replace the interatomic distances r_i, r_j, …, with the objects x_i, x_j, …, and replace the interatomic potential energies $u(r_i)$, $u(r_j)$, …, with the properties $h(x_i)$, $h(x_j)$, …. The only major change to the above formalism is in the covariance function in Eq. (2.6). In place of Eq. (2.7), we must write

$$K_{ij} = a\exp\left(-\frac{d\left(x_i, x_j\right)^2}{2L^2}\right), \tag{2.23}$$

where $d(x_i, x_j)$ measures the degree of similarity between objects x_i and x_j. The specific definition of $d(x_i, x_j)$ is arbitrary, however for excellent performance of Bayesian optimization it is essential that $d(x_i, x_j)$ be chosen after giving very careful consideration to the physics of the problem under study. Usually, $d(x_i, x_j)$ is defined as

$$d\left(x_i, x_j\right) = \left|\left|\phi(x_i) - \phi\left(x_j\right)\right|\right|, \tag{2.24}$$

where $\phi(x_i)$ and $\phi(x_j)$ are referred to as *feature vectors* (or *descriptors*) for the objects x_i and x_j, respectively (see Eq. (1.4) for the definition of the || || notation).

As discussed in the previous chapter, feature vectors are real-valued vectors which encode the key physics of the system of interest. In the example of the diatomic molecule, the feature vector for the distance r_i was simply set to $\phi(r_i) = r_i$. However, in most cases it is not so obvious which feature vectors one should use in order to achieve good performance with Bayesian optimization, and a great deal of physical intuition is needed to deduce such feature vectors. In any case, once such feature vectors are available, Bayesian optimization proceeds exactly as described in the sections above.

In this chapter, we have discussed Bayesian optimization in the context of minimization. However, Bayesian optimization can also be used to solve problems related to maximization as well. For the case where we wish to find the value of r which maximizes the value of some function u, the expected improvement in Eq. (2.18) must be re-written as

$$EI(r_\alpha) = E_f\left[\max\left(V(r_\alpha) - u_{\max}, 0\right)\right], \tag{2.25}$$

where u_{max} is the largest value of u in the sample data. Moreover, Eq. (2.19) must be replaced with

$$EI(r_\alpha) = \left(\mu^*_\alpha - u_{\max}\right)\Phi\left(\frac{\mu^*_\alpha - u_{\max}}{\sqrt{K^*_{\alpha\alpha}}}\right) + \sqrt{K^*_{\alpha\alpha}}\varphi\left(\frac{\mu^*_\alpha - u_{\max}}{\sqrt{K^*_{\alpha\alpha}}}\right) \tag{2.26}$$

Equation (2.26) can be proven by following similar steps to those shown in Appendix 2.2. Apart from the definition of the expected improvement, no changes to the theoretical framework developed above are necessary for solving maximization problems via Bayesian optimization.

2.4 *R* Code for Bayesian Optimization

One of the great advantages of Bayesian optimization is that it is relatively easy to implement in a computational environment. A program for calculating the expected improvement using an initial sample of Br-Br interatomic distances and the corresponding potential energies is available online at http://www.packwood.icems.kyoto-u.ac.jp/download/

This code is written in the *R* programming language, and can be executed within the *R* command line interface [3]. The *R* command line interface can be downloaded freely at https://www.r-project.org/, and numerous tutorials on *R* can be found online.

Successive applications of this code can be used to find the optimal distance between the Br atoms. Note that this code assumes that the energies have been pre-calculated for all points on a tight grid of bond distances, and that the initial sample is drawn randomly from this pre-calculated data. Obviously there is no need to perform Bayesian optimization in this case, as the equilibrium distance could be identified by directly looking at the pre-calculated data. In realistic applications of Bayesian optimization, the code will need to interface with first-principles calculation software or an experimental apparatus in order to obtain the sample data and subsequent measurements.

Appendix 2.1

To prove Eq. (2.12) – (2.14), we first substitute Eqs. (2.10) into (2.11) to obtain

$$f\left(v_\alpha, v_\beta, \ldots, v_\gamma \middle| u_i, u_j, \ldots, u_k\right) \propto g\left(u_i, u_j, \ldots, u_k \middle| v_\alpha, v_\beta, \ldots, v_\gamma\right) g\left(v_\alpha, v_\beta, \ldots, v_\gamma\right) \tag{2.27}$$

Because the likelihood function and the prior density are Gaussian probability densities, Eq. (2.27) simplifies to (by the basic properties of conditional densities [5])

$$f\left(v_\alpha, v_\beta, \ldots, v_\gamma \middle| u_i, u_j, \ldots, u_k\right) \propto g\left(v_\alpha, v_\beta, \ldots, v_\gamma, u_i, u_j, \ldots, u_k\right), \tag{2.28}$$

or, by using Eq. (2.8),

$$f\left(v_\alpha, v_\beta, \ldots, v_\gamma \middle| u_i, u_j, \ldots, u_k\right)$$
$$\propto \exp\left(-\frac{1}{2}\left[\begin{pmatrix} \mathbf{v}_{\alpha:\gamma} \\ \mathbf{v}_{i:k} \end{pmatrix} - \begin{pmatrix} \boldsymbol{\mu}_{\alpha:\gamma} \\ \boldsymbol{\mu}_{i:k} \end{pmatrix}\right]^T \begin{bmatrix} \mathbf{K}_{\alpha:\gamma,\alpha:\gamma} & \mathbf{K}_{\alpha:\gamma,i:k} \\ \mathbf{K}_{i:k,\alpha:\gamma} & \mathbf{K}_{i:k,i:k} \end{bmatrix}^{-1} \left[\begin{pmatrix} \mathbf{v}_{\alpha:\gamma} \\ \mathbf{v}_{i:k} \end{pmatrix} - \begin{pmatrix} \boldsymbol{\mu}_{\alpha:\gamma} \\ \boldsymbol{\mu}_{i:k} \end{pmatrix}\right]\right) \tag{2.29}$$

This expression can be simplified using an identity which applies to block matrices (see, Ref. [6])

$$\begin{bmatrix} \mathbf{K}_{\alpha:\gamma,\alpha:\gamma} & \mathbf{K}_{\alpha:\gamma,i:k} \\ \mathbf{K}_{i:k,\alpha:\gamma} & \mathbf{K}_{i:k,i:k} \end{bmatrix}^{-1}$$
$$= \begin{bmatrix} \left(\mathbf{K}_{\alpha:\gamma,\alpha:\gamma} - \mathbf{K}_{\alpha:\gamma,i:k}\mathbf{K}_{i:k,i:k}^{-1}\mathbf{K}_{i:k,\alpha:\gamma}\right)^{-1} & \left(\mathbf{K}_{\alpha:\gamma,\alpha:\gamma} - \mathbf{K}_{\alpha:\gamma,i:k}\mathbf{K}_{i:k,i:k}^{-1}\mathbf{K}_{i:k,\alpha:\gamma}\right)^{-1}\mathbf{K}_{\alpha:\gamma,i:k}\mathbf{K}_{i:k,i:k}^{-1} \\ -\left(\mathbf{K}_{i:k,i:k} - \mathbf{K}_{i:k,\alpha:\gamma}\mathbf{K}_{\alpha:\gamma,\alpha:\gamma}^{-1}\mathbf{K}_{\alpha:\gamma,i:k}\right)^{-1}\mathbf{K}_{i:k,\alpha:\gamma}\mathbf{K}_{\alpha:\gamma,\alpha:\gamma}^{-1} & \left(\mathbf{K}_{i:k,i:k} - \mathbf{K}_{i:k,\alpha:\gamma}\mathbf{K}_{\alpha:\gamma,\alpha:\gamma}^{-1}\mathbf{K}_{\alpha:\gamma,i:k}\right)^{-1} \end{bmatrix} \tag{2.30}$$

Substituting Eqs. (2.30) into (2.29) and performing some tedious but straightforward algebraic manipulations yields Eqs. (2.12)–(2.14).

Appendix 2.2

To prove Eq. (2.19), we write

$$\begin{aligned} EI(r_\alpha) &= E_f[\min(u_{\min} - V(r_\alpha), 0)] \\ &= \int_{-\infty}^{u_{\min}} (u_{\min} - z) \frac{1}{\sqrt{2\pi K^*_{\alpha\alpha}}} e^{-(z-\mu^*_\alpha)^2/2K^*_{\alpha\alpha}} dz \\ &= \underbrace{u_{\min} \int_{-\infty}^{u_{\min}} \frac{1}{\sqrt{2\pi K^*_{\alpha\alpha}}} e^{-(z-\mu^*_\alpha)^2/2K^*_{\alpha\alpha}} dz}_{A} - \underbrace{\int_{-\infty}^{u_{\min}} \frac{z}{\sqrt{2\pi K^*_{\alpha\alpha}}} e^{-(z-\mu^*_\alpha)^2/2K^*_{\alpha\alpha}} dz.}_{B} \end{aligned} \tag{2.31}$$

The term A on the right-hand side of Eq. (2.31) is equal to

$$A = u_{\min} \Phi\left(\frac{u_{\min} - \mu^*_\alpha}{\sqrt{K^*_{\alpha\alpha}}}\right), \tag{2.32}$$

by the definition of the cumulative normal distribution. As for the term B in Eq. (2.31), we write

$$\begin{aligned} B &= \int_{-\infty}^{u_{\min}} \frac{\mu^*_\alpha + (z - \mu^*_\alpha)}{\sqrt{2\pi K^*_{\alpha\alpha}}} e^{-(z-\mu^*_\alpha)^2/2K^*_{\alpha\alpha}} dz. \\ &= \mu^*_\alpha \Phi\left(\frac{u_{\min} - \mu^*_\alpha}{\sqrt{K^*_{\alpha\alpha}}}\right) + \underbrace{\int_{-\infty}^{u_{\min}} \frac{(z - \mu^*_\alpha)}{\sqrt{2\pi K^*_{\alpha\alpha}}} e^{-(z-\mu^*_\alpha)^2/2K^*_{\alpha\alpha}} dz.}_{C} \end{aligned} \tag{2.33}$$

By substituting the variable

$$h = \frac{z - \mu^*_\alpha}{\sqrt{2K^*_{\alpha\alpha}}} \tag{2.34}$$

into the term C in Eq. (2.33) and performing the integration, we obtain

$$\begin{aligned} C &= \left(\frac{2K^*_{\alpha\alpha}}{\pi}\right)^{1/2} \int_{-\infty}^{u_{\min}} he^{-h^2} dh \\ &= -\left(\frac{K^*_{\alpha\alpha}}{2\pi}\right)^{1/2} e^{-\left(u_{\max}-\mu^*_\alpha\right)^2/2K_{\alpha\alpha}} \\ &= -\sqrt{K^*_{\alpha\alpha}}\phi\left(\frac{u_{\min}-\mu^*_\alpha}{\sqrt{K^*_{\alpha\alpha}}}\right) \end{aligned} \tag{2.35}$$

where the definition of the standard normal probability density was used. We obtain the result after combining Eqs. (2.31), (2.32), (2.33) and (2.35).

Appendix 2.3

To prove Eqs. (2.20) and (2.21), we take the logarithm of the prior probability density in Eq. (2.5) to obtain

$$\begin{aligned} \log g\left(u_i, u_j, \ldots, u_k\right) &= -\frac{1}{2}(\mathbf{u}-\boldsymbol{\mu})^T(a\mathbf{R})^{-1}(\mathbf{u}-\boldsymbol{\mu}) - \frac{1}{2}\log|a\mathbf{R}| - \frac{s}{2}\log 2\pi \\ &= \underbrace{-\frac{1}{2}(\mathbf{u}-\boldsymbol{\mu})^T(a\mathbf{R})^{-1}(\mathbf{u}-\boldsymbol{\mu})}_{A} \underbrace{-\frac{s}{2}\log\left(a|\mathbf{R}|^{1/s}\right)}_{B} \underbrace{-\frac{s}{2}\log 2\pi}_{C}. \end{aligned} \tag{2.36}$$

In the first line of Eq. (2.36), we used the definition of the matrix $\mathbf{R}$ in Eq. (2.22) and the fact that $a\mathbf{R} = \boldsymbol{\Sigma}$. In the second line, we used the fact that $|a\mathbf{R}| = a^s|\mathbf{R}|$, which follows from the basic properties of determinants. Solving the equation $\partial \log g(u_i, u_j, \ldots, u_k)/\partial a = 0$ gives

$$a = \frac{1}{s}(\mathbf{u}-\boldsymbol{\mu})^T\mathbf{R}^{-1}(\mathbf{u}-\boldsymbol{\mu}), \tag{2.37}$$

which is simply Eq. (2.21). To obtain an expression for L, fist note that the term A reduces to

$$A = -s/2 \tag{2.38}$$

upon substituting Eq. (2.37). Substituting Eq. (2.37) into the term marked B gives

$$B = -\frac{s}{2}\log\left(\frac{1}{s}|\mathbf{R}|^{1/s}(\mathbf{u}-\boldsymbol{\mu})^T\mathbf{R}^{-1}(\mathbf{u}-\boldsymbol{\mu})\right). \tag{2.39}$$

Substituting Eqs. (2.38) and (2.39) into Eq. (2.34), and noting that the terms A and C are independent of a and L, gives Eq. (2.22). Finally, by noting that maximization of the logarithm of the prior probability density is equivalent to maximizing the prior probability density itself, we arrive at the result.

References

1. Rasmussen CE, Williams CKI. Gaussian processes for machine learning. MA, USA: The MIT Press; 2016 (Chapter 4).
2. Kresse G, Furthmuller J. Efficient iterative schemes for ab initio total energy calculations using a plane-wave basis set. Phys Rev B. 1996;54:11169–86.
3. R Core Team. R: a language and environment for statistical computing. R foundation for statistical computing. https://www.R-project.org/ (2017).
4. Frazier P, Wang J. Bayesian optimization for materials design. In: Lookman T, Alexander FJ, Rajan K, editors. Information science for materials discovery and design. Springer Series in Materials Science 225, Switzerland: Springer International Publishing; 2016.
5. Miller I, Miller M. John E. Freund's mathematical statistics with applications. 7th ed. Upper Saddle River, NJ, USA: Pearson Prentice-Hall; 2014.
6. Petersen KB, Pedersen MS. The matrix cookbook. http://matrixcookbook.com (Section 9.1.3). 15 Nov 2012.

Chapter 3
Bayesian Optimization of Molecules Adsorbed to Metal Surfaces

In the previous chapter, we saw how Bayesian optimization is implemented in practice by considering a diatomic molecule. Of course, there is little point in applying Bayesian optimization to such a simple system, as the full potential energy curve can be quickly calculated using simple quantum chemistry. In this chapter, we consider a more complex situation, consisting of organic molecules adsorbed to a metal surface.

3.1 Density Functional Theory for Surface Science

Many treatments on computational materials science are quick to remark on the great improvements in computational power made over the last decades. This increases in computational power has allowed for first-principles calculations for large systems to be performed in tandem with experiments, and such calculations are now routinely reported in experimental materials science research papers. Of the available first-principles methodologies, density functional theory (DFT) is arguably the preferred method for much of computational materials science, and the development of novel exchange-correlation functionals and dispersion corrections have significantly broadened the applicability of DFT in materials science research [1–3]. However, despite the increasingly widespread adoption of DFT in materials science, there still remains an incredible number of systems that are extremely difficult to study via DFT, due to the share number of atoms needed to model such systems as well as their complicated potential energy landscapes.

In this chapter, we focus on the case of metal surfaces possessing organic molecule adsorbates. While metal surfaces are regularly modified with organic adsorbates in experimental research, DFT-based structure optimizations for organic molecules on a surface require prohibitively long computational times. Such calculations can be broken into two steps: (a) choice of adsorption sites and orientations for each molecule, and (b) structural relaxation of the system to find the

D. Packwood, *Bayesian Optimization for Materials Science*,
SpringerBriefs in the Mathematics of Materials 3,
https://doi.org/10.1007/978-981-10-6781-5_3

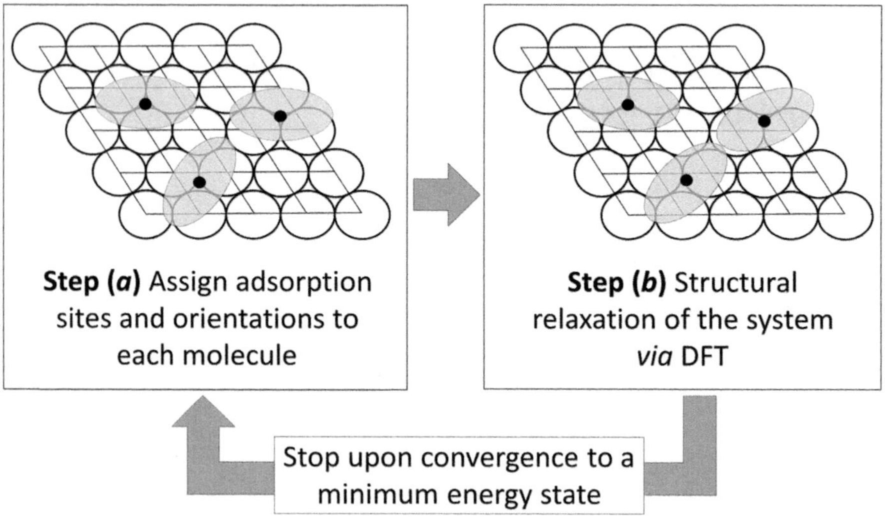

Fig. 3.1 Scheme for identifying the energetically optimal conformation of molecules adsorbed to a surface via density functional theory (DFT) calculations. The black circles show the atoms of a (111) surface, and the grey ovals represent adsorbed molecules, and the black points represent the center-of-mass of the molecules. The black points identify the adsorption site of the molecule on the surface

energy minimizing conformation of the molecule (Fig. 3.1). Steps (a) and (b) are then repeated until convergence to a global minimum energy structure is obtained. Step (b) may involve around 1000 atoms and demand weeks of computation if performed via DFT. On the other hand, there are usually very many ways of placing the molecules on the surface, meaning that there are very many ways of performing Step (a). In other words, in order to find the global energy minimizing state of the system, Step (b) may need to be repeated a very large number of times. Consequently, months or years of computational time may be necessary until convergence to a global minimum structure is obtained. The situation is particularly hopeless when large numbers of molecules are considered, because the number of ways of performing Step (a) increases exponentially with the number of molecules under consideration.

In order to use DFT to compute the energy minimizing conformation and arrangement of molecules on a metal surface, it is necessary to reduce the number of times that Step (*b*) in the above scheme must be repeated. In other words, we need a method which can choose the adsorption sites and orientations for the molecules in Step (*a*) in such a way that very few repetitions the scheme in Fig. 3.1 are needed to find the energy minimum conformation of the system.

3.2 Bayesian Optimization for Surface Science

Bayesian optimization is one candidate for reducing the computational load for structure prediction for molecules adsorbed metal surfaces. Groups in Europe, Japan, and possibly elsewhere are currently applying Bayesian optimizing for this purpose, and while this field is in its infancy, early results are very promising. Recently, Todorovic, Rinke, and co-workers reported on the use of Bayesian optimization to find the energetically optimal position and orientation of single molecules adsorbed metal surfaces [4]. The resulting code, which they call BOSS (Bayesian Optimisation Structure Search), is shown to succeed within tens of iterations of the Bayesian optimization procedure for a variety of adsorbate molecules and surfaces. The performance reported by this group is particularly outstanding when compared to uniform random sampling of positions and orientations for the molecule adsorbates, which is expected to require hundreds of DFT calculations until the energy minimizing conformation is found.

Recently, we applied Bayesian optimization to predict the energy-minimum arrangement of two medium-sized organic molecules adsorbed to a metal surface, and similarly found that Bayesian optimization could succeed within tens of DFT calculations [5]. Specifically, we considered two dibromo-bianthracene molecules (10,10′-dibromo-9,9′-bianthracene, or Br_2BA) adsorbed to a copper (111) surface (Fig. 3.2a). This molecule consists of two anthracene moieties connected together by a single C-C bond. Readers familiar with Br_2BA may associate it with graphene

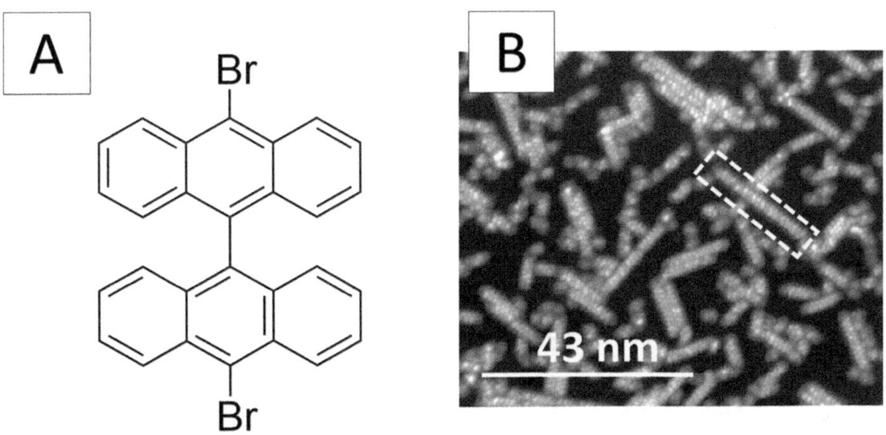

Fig. 3.2 **a** Chemical structure of the 10,10′-dibromo-9,9′-bianthracene (Br_2BA) molecule. **b** Scanning tunneling microscopy (STM) image a Cu(111) surface following deposition of Br_2BA. The Br_2BA molecules assemble into chain-like structures, one of which is identified by the dotted white box. STM conditions: voltage 1.1 V, tunneling current 10 pA, imaging temperature 5.6 K, annealing temperature 400 °C. STM image acquired by Patrick Han (AIMR, Tohoku University), using an STM created by Taro Hitosugi (Tokyo Institute of Technology) and co-workers

nanoribbon fabrication [6–9]. However, our interest in Br_2BA is mainly that it undergoes a simple self-assembly process when deposited on Cu(111), resulting in the formation of (mainly) linear chains of Br_2BA molecules (Fig. 3.2b). For the case of two Br_2BA molecules adsorbed to Cu(111), we should therefore find that the global energy minimum of the system corresponds to the two molecules are aligned in a chain-like fashion. In this remainder of this chapter, we will review this study and explain in detail at how Bayesian optimization was implemented.

3.2.1 Preliminary Computational Study

Before jumping into the situation of two Br_2BA molecules on Cu(111), we first performed a preliminary computational study for a single Br_2BA molecule on Cu (111). As we show here, the purpose of such a preliminary computational study is to find shortcuts for performing both steps (a) and (b) in Fig. 3.1.

Following the DFT methodology described in [10], a single Br_2BA molecule was placed in several positions and orientations on a perfect Cu(111) surface, and a local structure relaxation for the molecule performed for each case. The displacement of each atom in the molecule following structure relaxation was then averaged over all cases, resulting in the 'averaged' adsorption conformation shown in Fig. 3.3. Strong van der Waals interactions with the surface mean that one end of each anthracene unit lies nearly parallel to the surface. The other end of the anthracene unit bends away from the surface, which decreases steric repulsions between protons. The strong van der Waals interactions result in very strong adsorption energies (in the

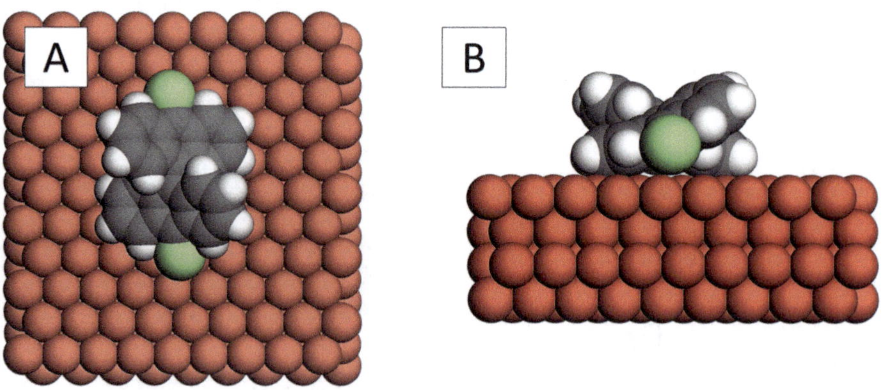

Fig. 3.3 Conformation of a single Br_2BA molecule adsorbed to a Cu(111) surface, as calculated using a combination of DFT and an 'averaging procedure reported' in [10]. **a** Shows the conformation of the molecule viewed with the Cu(111) surface in the place of the page, and **b** Shows the conformation with the Cu(111) surface perpendicular to the page. Red-brown, white, gray, and green atoms correspond to copper, hydrogen, carbon, and bromine atoms, respectively

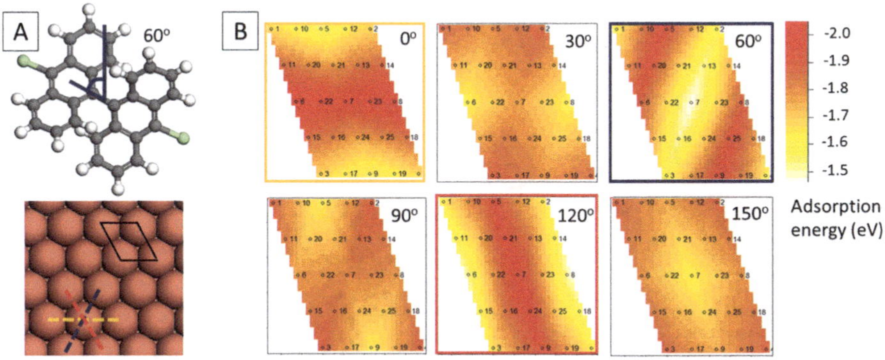

Fig. 3.4 **a** A single Br_2BA molecule adsorbed to a Cu(111) in the '60°' orientation (top). The conformation of the molecule is identical to the conformation showed in Fig. 3.3. The orientation of the molecule is determined according to the angle formed between the central C–C bond and the vertical direction. The Cu(111) surface is drawn at the bottom of (**a**). A single unit cell and the directions of the lattice places are indicated by the black box and colored dotted lines, respectively. **b** Adsorption energy maps, calculated by scanning the molecule conformation in (**a**) above a unit cell of the Cu(111) surface in various orientations. In these diagrams the energy at a specific point corresponds to the adsorption energy when the center-of-mass of the molecule lies directly above that point. The points are numbered for convenience, however this numbering has no specific meaning. Images plotted with the AKIMA package [11, 12]

order of 2 eV), suggesting that interactions between Br_2BA molecules on Cu(111) (expected to be in the order of a few tenths of an electron volt) would have little effect on the conformation of the adsorbed molecules.

To determine which adsorption sites and orientations should be sampled during Step (a) of the cycle in Fig. 3.1, the averaged adsorption conformation from Fig. 3.3 was scanned across a single unit cell of Cu(111) in a variety of orientations, and the adsorption energy of the molecule was calculated on-the-fly via DFT calculations [10]. This resulted in the adsorption energy maps shown in Fig. 3.4, from which we can identify the points in the dark red trenches being the preferred adsorption sites of the molecule. These adsorption energy maps clearly show a preference for orientations which point in the direction of the Cu(111) lattice planes. To implement Step (a) of Fig. 3.1, it is therefore sufficient to consider only these low-energy adsorption sites and orientations when deciding where to place the two molecules. Finally, to check the effect of the intermolecular interaction on the conformation of surface-adsorbed Br_2BA molecules, a small number of structural relaxations were performed using two adsorbed Br_2BA molecules sitting close proximity to each other. All calculations started from the averaged conformation showed in Fig. 3.3, and very negligible distortion to this conformation was observed during the course of the structural relaxation. This supports the assumption that the surface has the dominant effect on the conformation of single Br_2BA molecules adsorbed to Cu (111), and also means that we can drop the structural optimization part of Step (b) in Fig. 3.1 and perform a static energy calculation instead.

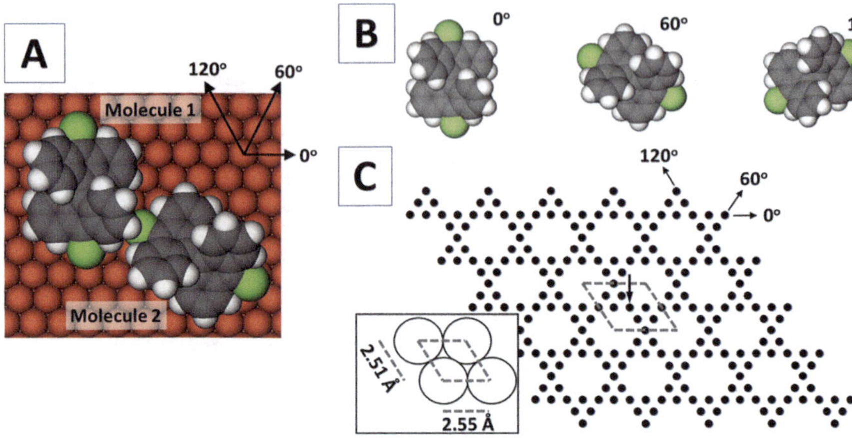

Fig. 3.5 Illustration of the specific optimization problem considered here. **a** shows two Br_2BA molecules (marked molecule 1 and molecule 2) adsorbed to a Cu(111) surface. The direction of the three lattice planes is marked. **b** Br_2BA molecules in the 0°, 60°, and 120° orientations, respectively. **c** Possible adsorption sites for molecule 2, when the center-of-mass of molecule 1 is fixed at the adsorption site marked by the down-pointing arrow in the 0° orientation. Rows of adsorption sites that lie in the direction of the arrow marked by 0° (respectively 60°, 120°) permit the molecule to adsorb in the 0° (respectively 60°, 120°) orientation. A single unit cell of the Cu (111) surface is marked by the dotted grey box and is illustrated by the insert. Figure from reference [5]. Copyright 2017, The Japan Society of Applied Physics

3.2.2 *Statement of Optimization Problem*

The specific optimization problem is shown in Fig. 3.5. One molecule ('molecule 1') sits in the adsorption site marked by the down-pointing arrow, and the position of the other molecule ('molecule 2') can be varied. Molecule 2 can only reside on the adsorption sites shown in Fig. 3.5, and can only adopt one of the three orientations shown there. We define a *molecule arrangement* as a choice of one adsorption site and orientation. Let σ denote a single molecule arrangement. The *stabilization energy* for molecule arrangement σ is defined as

$$\varepsilon(\sigma) = u_1 + u_2 + v_{12}, \tag{3.1}$$

where u_1 is the energy of interaction between molecule 1 and the surface, u_2 is the energy of interaction between molecule 2 and the surface, and v_{12} is the energy of interaction between molecule 1 and molecule 2. Note that u_1 and u_2 are equivalent to the adsorption energy of molecules 1 and 2 on the surface, respectively. Our goal is to find the molecule arrangement with the minimum stabilization energy, i.e., the molecule arrangement σ which gives the most negative value of $\varepsilon(\sigma)$, using Bayesian optimization.

While each of the terms in Eq. (3.1) can be calculated via routine DFT, these calculations can demand up to 30 h of computation on our hardware. It is therefore important to identify the optimum molecule arrangement by checking as few molecule arrangements as possible. To compute Eq. (3.1) via DFT, we actually use the following equation,

$$\varepsilon(\sigma) = E(\sigma) - E_s(\sigma) - E_{m1}(\sigma) - E_{m2}(\sigma), \tag{3.2}$$

where $E(\sigma)$ is the total energy of the system (with both molecules on surface), $E_s(\sigma)$ is the energy of the surface alone (with both molecules removed from surface), and $E_{m1}(\sigma)$ and $E_{m2}(\sigma)$ are respectively the energies of molecule 1 and molecule 2 alone. Because molecules 1 and 2 have identical conformations, $E_{m1}(\sigma) = E_{m2}(\sigma)$. While $E_{m1}(\sigma)$ and $E_{m2}(\sigma)$ can be calculated within tens of minutes to reasonable accuracy, the calculations of $E(\sigma)$ and $E_s(\sigma)$ typically involve between 400 and 800 atoms and require considerably longer computational times. The large variation in the number of atoms is due to the fact that, in some molecule arrangements, molecule 1 and 2 are quite widely separated on the surface, which requires larger simulation boxes and hence larger numbers of atoms in the calculation.

3.2.3 *Data Description*

The data used in this study comprises of all possible molecule arrangement in which the minimum interatomic distance between molecules 1 and 2 is between 1 and 4 Å. Molecule arrangements in which the minimum interatomic distance is less than 1 Å are expected to be very unstable, as the van der Waals radii of the two molecules will strongly overlap. On the other hand, we do not expect for the molecules in the optimum molecule arrangement to be separated by more than 4 Å, because at these distances the atoms in molecule 2 are not expected to feel a strong attractive force from the presence of molecule 1. This criterion resulted in 480 different molecule arrangements to consider (210 for the case where molecule 2 is in the 0° orientation, 160 for the case where molecule 2 is in the 60° orientation, and 110 for the case where molecule 2 is in the 120° orientation).

Even after reducing the number of molecule arrangements to 480 possibilities, there remain many molecule arrangements which are obviously unstable (Fig. 3.6). Whereas the optimal molecule arrangement is expected to have a stabilization energy in the order of −0.1 eV, the unphysical molecule arrangements are expected to have stabilization energies in the order of +10 to +100 eV. Because of the overwhelming magnitude of the stabilization energies of the unphysical molecule arrangements, the mean of the posterior distribution for the stabilization energies (as calculated by Eq. 2.16) will take on predominantly positive values, and may not be negative for metastable molecule arrangements. In turn, this will make detection of the optimal molecule arrangement via Bayesian optimization extremely difficult. In order to efficiently identify the optimal molecule arrangement via Bayesian optimization, it is therefore preferable to remove these unphysical molecule arrangements from the calculation of the posterior distribution.

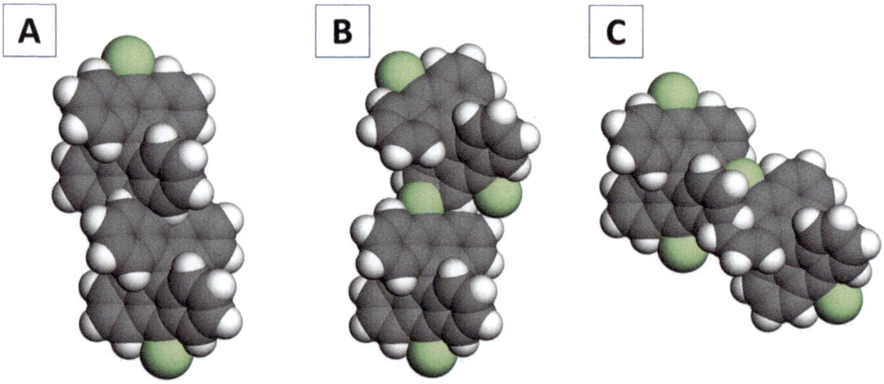

Fig. 3.6 Example of three 'unphysical' molecule arrangements (see text). In each case, the molecules are too close together. This results is very large and positive stabilization energies [see Eq. (3.2)]

Thankfully, the unphysical molecule arrangements can be easily identified by a DFT-based screening procedure. Here, we neglect dispersion interactions and the Cu atoms of the surface, and run DFT energy calculations for each of the 480 molecule arrangements. Then, molecule arrangements with stabilization energies exceeding 0.15 eV are identified as being unphysical. The neglect of dispersion interactions and the surface atoms is acceptable for this screening procedure, because for the case of the unphysical molecule arrangements the stabilization energy is dominated by short-range electrostatic repulsions between atoms of the molecules. Following this screening procedure, which only requires a couple of days of computational time on modern hardware, we identified 186 of the 480 molecule arrangements as being unphysical (thus, 294 molecule arrangements are identified as 'physical'). These molecule arrangements were neglected in the calculation of the posterior distributions described in the following sections.

3.2.4 Choice of Feature Vectors

As emphasized in Chap. 1, the choice of feature vectors a critical influence on the performance of Bayesian optimization. To the best of our knowledge there are no reports of feature vectors which are specifically designed for interacting molecules. However, plenty of effort has been made in designing feature vectors for single molecules, particular for the purpose of predicting atomization energies [13, 14].

The *Coulomb matrix* is a popular feature vector for single molecules [13]. In the present study, we constructed an *interaction Coulomb matrix*, which is defined as a matrix $\mathbf{c}(\sigma) = [c_{ij}(\sigma)]_{n\times n}$. Here, n is the number of atoms per molecule, the rows correspond to the atoms of molecule 1, the columns correspond to the atoms of molecule 2, and

$$c_{ij}(\sigma) = \frac{q_i q_j}{|\mathbf{r}_i - \mathbf{r}_j|}, \tag{3.3}$$

where q_i and q_j are the atomic numbers of atoms i and j, respectively, and $|\mathbf{r}_i - \mathbf{r}_j|$ is the distance between atoms i and j. As a feature vector for molecule arrangement σ, we choose

$$\phi(\sigma) = (\mathbf{c}_{1,}(\sigma), \mathbf{c}_{2,}(\sigma), \ldots, \mathbf{c}_{n,}(\sigma)), \tag{3.4}$$

where $\mathbf{c}_j(\sigma)$ is the jth row of the interaction Coulomb matrix. The interaction Coulomb matrix essentially describes the molecules as a cloud of point charges interacting through Coulomb interactions, and does not directly describe the effects of exchange interactions or electron correlation between molecules. Moreover, the interaction Coulomb matrix does not describe the surface-molecule interaction. However, despite not capturing these important features, Bayesian optimization using the Coulomb matrix performs very well for the present problem (see Sects. 2.5 and 2.6). We will return to this point in Sect. 2.7.

3.2.5 Training of Hyperparameters

In the previous chapter, we introduced the marginal likelihood technique for choosing the hyperparameters (a and L in Eq. 2.7). While the marginal likelihood technique is certainly very helpful, there is no theoretical guarantee that it will pick the most appropriate values of the hyperparameters for the specific problem under study. In general, one should always start from physical considerations when deciding upon the hyperparameter values.

In the present case, we decide upon the hyperparameter values by considering a 'gas-phase' system, in which the Cu atoms of the copper substrate are removed from the DFT calculations. With the Cu atoms removed, the calculation of the stabilization energy of a molecule arrangement (with a van der Waals exchange-correlation functional [2]) is relatively quick (around 20–20 min), and the stabilization energy of all 294 molecule arrangements can be computed within a few days and stored as a database. Bayesian optimization is then performed on the gas-phase molecule arrangements as described in Chap. 2, however instead of performing a new DFT calculation at the end of each iteration, we simply retrieve the appropriate stabilization energy from the database and add it to our sample data. With this approach, we can easily perform Bayesian optimization for a variety of values of the parameters a and L, and identify a parameter regime in which the optimal molecule arrangement tends to be identified relatively quickly. Moreover, by checking against the minimal stabilization energy in the database, we can easily confirm the convergence to the optimal molecule arrangement.

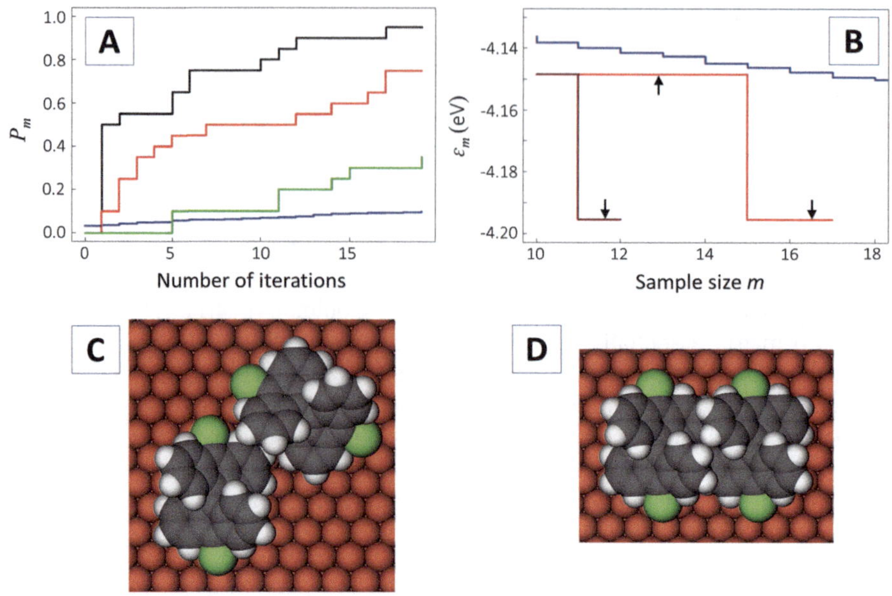

Fig. 3.7 **a** Training of the Bayesian optimization procedure by excluding the Cu atoms from the density functional theory (DFT) energy calculations. P is the probability that the global optimum molecule arrangement has been identified after a given number of iterations of the Bayesian optimization procedure. The green (respectively red, black) lines indicate calculations with $a = 1$ and $b = 10^{-3}$ (respectively $b = 7 \times 10^{-5}$, $b = 2 \times 10^{-5}$) in the covariance matrix, where $b = 1/(2L^2)$. The blue line represents the case where uniform random sampling is used instead of Bayesian optimization. P was estimated from 20 repetitions of the Bayesian optimization procedure (2000 times for the uniform random sampling case). **b** Bayesian optimization applied to the full system with Cu atoms included, using $a = 1$ and $b = 7 \times 10^{-5}$ in the covariance matrix. ε_m is the minimum stabilization energy in the sample data. Sample size (m) is equivalent to the number of evaluations of Eq. (3.2) via DFT. The red and dark-red lines indicate two independent runs of the procedure. The blue line is the lower-bound to ε_m when uniform random sampling is applied this system. **c**, **d** The molecule arrangements corresponding to the stabilization energy marked up-pointing arrow and down-pointing arrows in (**b**), respectively. Figure taken from reference [5]. Copyright 2017, The Japan Society of Physics

Figure 3.7a shows the probability of detecting the optimal gas-phase molecule arrangement for a variety of values of parameters a and $b = 1/(2L^2)$, as a function of the number of iterations of Bayesian optimization. In each case, initial samples of 10 molecule arrangements were used. Excellent performance for the choice $a = 1$ and b between 1×10^{-5} and 1×10^{-4} was observed. For the choice of $a = 1$ and $b = 7 \times 10^{-5}$, the optimal molecule arrangement was identified within 15 iterations of Bayesian optimization. This performance is particularly spectacular when compared to the blue line in Fig. 3.7a, which shows the probability of identifying the optimal molecule arrangement via simple random sampling of the molecule arrangements.

The calculations above suggest using $a = 1$ and b between 1×10^{-5} and 1×10^{-4} for Bayesian optimization of the full system (with copper atoms included). By using these values of the hyperparameters, we are implicitly assuming that the performance of Bayesian optimization is not strongly affected by the presence of the surface. The validity of this assumption is unclear at present, and further studies into this problem are necessary in order to improve the application of machine learning methods to modified surfaces.

3.2.6 Predictive Performance

Figure 3.7b shows the results of Bayesian optimization the full system (with copper atoms included), using the hyperparameter values of $a = 1$ and $b = 7 \times 10^{-5}$ [where $b = 1/(2L^2)$] and initial samples of 10 random molecule arrangements. In Fig. 3.7b, ε_m is the minimum stabilization energy in the sample. In two independent trials of the Bayesian optimization procedure, ε_m reached a value of around −4.94 eV within only 1 and 5 iterations of the Bayesian optimization procedure, respectively. This rapid convergence may be a lucky result, because in both cases the initial samples contained the same low-energy molecule arrangement (Fig. 3.7c), and this low-energy molecule arrangement may be very important for predicting the optimal molecule arrangement. While it is not possible to unambiguously confirm convergence of the stabilization energy, some additional reasoning suggests that the optimal molecule arrangement detected here (Fig. 3.7d) is in fact the true optimal molecule arrangement for this system. Firstly, it is well known that Br_2BA molecules adsorbed to a Cu(111) surface align in a chain-like fashion such as shown in Fig. 3.7d (also see Fig. 3.2b). It is therefore reasonable to assume that the structure in Fig. 3.7d lies in the region of the global stabilization energy minimum of the system. Secondly, additional DFT calculations for various chain-like alignments show that the prediction in Fig. 3.7d does in fact correspond to the true stabilization energy minimum of the system (Fig. 3.8).

While it is not possible to unambiguously compare the performance of Bayesian optimization to simple random sampling for the case of the full system, we can estimate a lower-bound to the minimal energy ε_m predicted by simple random sampling. This lower-bound is represented by the blue line in Fig. 3.7b, and was calculated by performing simple random sampling on the 'gas-phase' system (discussed in the previous section) and adding −4 eV to the stabilization energies (since the adsorption energy for a single Br_2BA molecule at any of the sites shown in Fig. 3.4 is around 2 eV). The blue line is a lower-bound to the stabilization energy, because the presence of charge cushions around the molecules, which are expected to contribute around +0.01 to +0.02 eV to the stabilization energies [10], have been ignored. These charge cushions result from displacement of charge from the surface upon adsorption of the molecules [15, 16]. Thus, our results demonstrate that Bayesian optimization achieves a superior performance compared to simple random sampling of molecule arrangements.

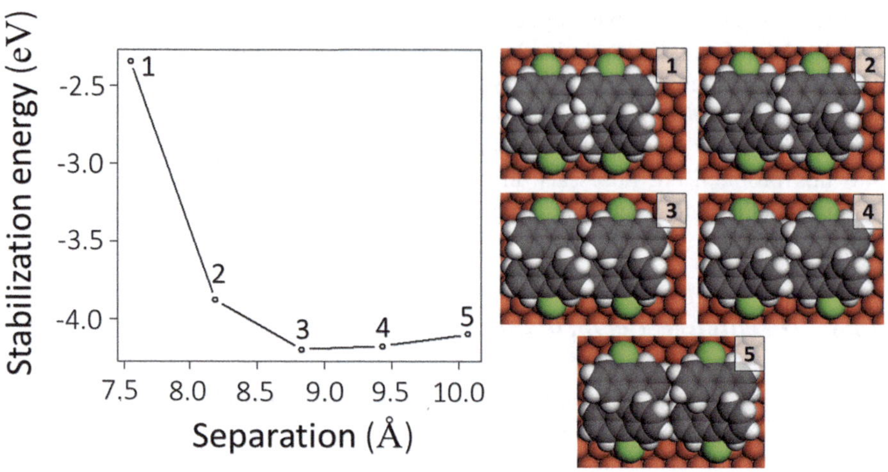

Fig. 3.8 Stabilization energies computed as a function of intermolecular separation for two Br_2BA molecules aligned in the direction of their anthracene units. The molecules are adsorbed to Cu(1111). Intermolecular separation is defined as the distance between the center of masses of the molecules. Stabilization energy was computed according to Eq. (3.2) and DFT. The images marked 1–5 on the right-hand side show the positions of the molecules at each intermolecular separation. Image 3 corresponds to the optimum molecule arrangement identified with Bayesian optimization [see Fig. 3.6d]. Figure from reference [5]. Copyright 2017, The Japan Society for Applied Physics

3.2.7 Discussion

The results shown above demonstrate that Bayesian optimization may be of great value for structure prediction problems in materials science. As stressed in Chap. 1 of this monograph, the success of Bayesian optimization is related to the fact that it uses 'global' information from all of the molecule arrangements in the sample when predicting the optimal molecule arrangement. To appreciate this point, compare the molecule arrangement shown in Fig. 3.7c (which corresponds to a local minimum) to the one shown in Fig. 3.7d (which corresponds to a global minimum). A classical gradient-based optimizer would have enormous trouble moving from this local minimum molecule arrangement to the global minimum, due to the presence of large barriers in the potential energy landscape between these two molecule arrangements. On the other hand, Bayesian optimization does not appear to be affected from this kind of trouble; once the sample data contains sufficient information to correlate the arrangement of molecules with the stabilization energy, it can quickly identify the optimal molecule arrangement regardless of where it lies in the potential energy landscape.

The calculations reported above assumed frozen internal degrees of freedom for both the molecule and surface. While this assumption is very reasonable for the specific system studied here (see the discussion in [5]), it is not necessary to employ it when applying Bayesian optimization. If relaxation of internal degrees of freedom was allowed in the present case, then Bayesian optimization would simply predict the adsorption site and orientation for molecule 2 which results in the lowest stabilization energy after relaxation of internal degrees of freedom. Such a calculation would resemble the one described in Sect. 1.2.3, Chap. 1. In general, if one intends to apply Bayesian optimization to a complex material, then the enumeration of possible starting configurations (molecule arrangements in the current study) should be carefully considered so that all minima in the potential energy landscape have chance of being reached following relaxation of the system.

As mentioned earlier, the feature vector chosen here (the interaction Coulomb matrix) essentially describes the molecule as a cloud of charges interacting via Coulomb repulsions. While this feature vector does not directly describe the surface-molecule interaction, or the exchange interactions that could take place between molecules, it was nonetheless sufficient for the Bayesian optimization procedure to quickly predict the optimal molecule arrangement from sample data. In the present case, this efficiency probably results from two facts. Firstly, the surface-molecule interaction energy varies little between the adsorption sites and orientations shown in Fig. 3.5 (for each of the adsorption sites and orientations shown in Fig. 3.5, the surface-molecule interaction is close to 2 eV for each case). Ignoring the effects such as charge cushion repulsions mentioned above (which may be significant in some systems), the effect of the surface is essentially to add a constant term to the stabilization energy. Secondly, the energies associated with exchange and correlation effects depend upon the distances between electrons in the molecules, and this information may be indirectly accounted for via the denominators of the interaction Coulomb matrix elements (Eq. 3.3). Thus, even though the interaction Coulomb matrix does not directly describe every effect that determines the stabilization energy of the molecule arrangements, it appears to contain sufficient information for Bayesian optimization to perform efficiently. A useful target for future research would therefore be to create a systematic guideline for necessary for a 'good' feature vector for a given type of system.

References

1. Perdew JP, Burke K, Ernzerhof M. Generalized gradient approximation made simple. Phys Rev Lett. 1997;78:1396.
2. Hamada I. van der Waals density functional made accurate. Phys Rev B. 2014;89:121103.
3. Tkatchenko A, Scheffler M. Accurate molecular van der Waals interactions from ground-state electron density and free-atom reference data. Phys Rev Lett. 2009;102:073005.
4. Todorovic M, Gutmann MU, Corander J, Rinke P. arXiv:1708.09274.
5. Packwood DM, Hitosugi T. Rapid prediction of molecule arrangements on metal surfaces *via* Bayesian optimization. Appl Phys Express. 2017;10:065502.
6. Cai J. Atomically precise bottom-up fabrication of graphene nanoribbons. Nature. 2010;466:470.
7. Han P, et al. Bottom-up graphene-nanoribbon fabrication reveals chiral edges and enantioselectivity. ACS Nano. 2014;8:9181.

8. Han P, et al. Self-assembly strategy for fabricating connected graphene nanoribbons. ACS Nano. 2015;9:12035.
9. Ruffieux P, et al. On-surface synthesis of graphene nanoribbons with zigzag edge topology. Nature. 2016;531:489.
10. Packwood DM, Han P, Hitosugi T. Chemical and entropic control on the molecular self-assembly process. Nat Commun. 2017;8:14463.
11. Akima H, Gabhardt A. Akima: interpolation of irregularly and regularly spaced data. R package version 0.5–12. 2015. http://CRAN.R-project.org/package=akima.
12. R Core Team. R: a language and environment for statistical computing. R Foundation for Statistical Computing. 2017. https://www.R-project.org/.
13. Rupp M, et al. Fast and accurate modeling of molecular atomization energies with machine learning. Phys Rev Lett. 2012;108:058301.
14. Hansen K, et al. Assessment and validation of machine learning methods for predicting molecular atomization energies. J Chem Theory Comput. 2013;9:3404.
15. Bagus PS, Germann K, Woll C. The interaction of C_6H_6 and C_6H_{12} with noble metal surfaces: electronic level alignment and the origin of the interface dipole. J Chem Phys. 2005;123:183109.
16. Witte G, et al. Vacuum level alignment at organic/metal junctions: "Cushion" effect and the interface dipole. Appl Phys Lett. 2015;87:263502.

GPSR Compliance

The European Union's (EU) General Product Safety Regulation (GPSR) is a set of rules that requires consumer products to be safe and our obligations to ensure this.

If you have any concerns about our products, you can contact us on ProductSafety@springernature.com

In case Publisher is established outside the EU, the EU authorized representative is:

Springer Nature Customer Service Center GmbH
Europaplatz 3
69115 Heidelberg, Germany

Batch number: 10704713

Printed by Printforce, the Netherlands